YOUTH CREATED MEDIA ON THE CLIMATE CRISIS

This timely book provides effective methods and authentic examples of teaching about climate change through digital and multimodal media production in the English Language Arts classroom. The chapters in this edited volume demonstrate the benefits of addressing climate change in the classroom through innovative media production and cover a range of different types of media, including video/digital storytelling, social media, art, music, and writing, with rich resources for instructions in every chapter.

Through the engaging ideas and strategies, the contributors equip educators with the critical tools for supporting students' media production. In so doing, they offer new perspectives on how students can employ media and production techniques to critique the status quo, call for change, and acquire new literacy skills. As the effects of the climate crisis become increasingly visible to the youth population, this book helps foster and support youth agency and activism. *Youth Created Media on the Climate Crisis: Hear Our Voices* is a necessary text for students, preservice teachers, and educators in literacy education, media studies, social and environmental studies, and STEM education.

Richard Beach is the Professor Emeritus at the University of Minnesota, USA. He is the author of numerous books, including *Teaching Climate Change to Adolescents: Reading, Writing, and Making a Difference; Teaching Literature to Adolescents, 4th Edition; Teaching Language as Action; Languaging Relations for Transforming Literacy and the Language Arts Classroom;* and *Teaching to Exceed the English Language Arts Common Core State Standards, 3rd Edition.*

Blaine E. Smith is an Associate Professor of the Practice in the Department of Teaching and Learning at Vanderbilt University, USA. Her scholarship is focused on multilingual adolescents' digital literacies and developing strategies for supporting teachers' integration of technology in diverse classrooms.

YOUTH CREATED MEDIA ON THE CLIMATE CRISIS

Hear Our Voices

Edited by Richard Beach and Blaine E. Smith

Routledge
Taylor & Francis Group

NEW YORK AND LONDON

First published 2024
by Routledge
605 Third Avenue, New York, NY 10158

and by Routledge
4 Park Square, Milton Park, Abingdon, Oxon, OX14 4RN

Routledge is an imprint of the Taylor & Francis Group, an informa business

ISBN: 978-1-032-37090-3 (hbk)
ISBN: 978-1-032-36900-6 (pbk)
ISBN: 978-1-003-33527-6 (ebk)
ISBN: 978-1-032-52288-3 (eBook+)

DOI: 10.4324/9781003335276

Typeset in Bembo
by KnowledgeWorks Global Ltd.

CONTENTS

YOUTH CLIMATE CONTRIBUTORS LIST

Julian Arenas is an Undergraduate Student at Williams College. julian.arenas031@gmail.com

Kyle Bartlett is a Doctoral Student in Music Learning and Teaching, Arizona State University. kbartle7@asu.edu

Linda Buturian is a Senior Teaching Specialist, Department of Curriculum and Instruction, University of Minnesota. butur001@umn.edu

Ardra Charath is an Undergraduate Student at the California Institute of Technology. ardracharath1@gmail.com

Amy Cutter-Mackenzie-Knowles is a Professor of Sustainability, Environment and Education at Southern Cross University, Australia. amy.cutter-mackenzie@scu.edu.au

Beth Ferguson is an Assistant Professor of Design at the University of California, Davis. bferguson@ucdavis.edu

Steve Goodman is the Founding Executive Director Emeritus of the Educational Video Center in New York City. stevegoodmanco@gmail.com

Nataliia Goshylyk is a Lecturer at the University of California, Berkeley, and an Associate Professor of the English Philology Department at Vasyl Stefanyk Precarpathian National University, Ukraine. ngoshylyk@berkeley.edu

Shiyan Jiang is an Assistant Professor of Learning Design and Technology at North Carolina State University. sjiang24@ncsu.edu

Michelle Jordan is an Associate Professor in the Mary Lou Fulton Teachers College at Arizona State University. michelle.e.jordan@asu.edu

Catherine Lockmiller is a Diversity Fellow and Assistant Health Science Librarian for Cline Library at Northern Arizona University. Catherine.Lockmiller@nau.edu

Antonio López is a Professor of Communications and Media Studies at John Cabot University in Rome, Italy. alopez@johncabot.edu

N. Claire Napawan is an Associate Professor of Landscape Architecture and the Landscape Architecture + Environmental Design Program Director at the University of California, Davis. ncnapawan@ucdavis.edu

Marek Oziewicz is a Professor of Literacy Education, Department of Curriculum & Instruction, University of Minnesota. mco@umn.edu

Emily Polk is an Advanced Lecturer in the Program in Writing and Rhetoric and the Coordinator for the Notation in Science Communication, Stanford University. empolk@stanford.edu

David Rousell is a Senior Lecturer in Creative Education at Royal Melbourne Institute of Technology, Australia. david.rousell@rmit.edu.au

Ji Shen is a Professor of STEM Education, Department of Teaching and Learning, University of Miami. j.shen@miami.edu

Blaine E. Smith is an Associate Professor, Department of Teaching and Learning, Vanderbilt University. blaine.smith@vanderbilt.edu

Brett L. Snyder is an Associate Professor of Design at the University of California, Davis. brett@chengsnyder.com

Scott Spicer is a Media Outreach Librarian and Head of Libraries Media Services, University of Minnesota Libraries–Twin Cities. spic0016@umn.edu

Evan Tobias is an Associate Professor of Music Learning and Teaching at Arizona State University. etobias@asu.edu

Thilinika Wijesinghe is currently a PhD candidate, Faculty of Education, Southern Cross University, Australia. thilinika.wijesinghe@scu.edu.au

Liane Xu is a Student at the Massachusetts Institute of Technology. lianexu@mit.edu

Steven J. Zuiker is an Associate Professor, Mary Lou Fulton Teachers College, Arizona State University. steven.zuiker@asu.edu

INTRODUCTION

Need for This Book

Blaine E. Smith and Richard Beach

From 1901 to 2020

- Global temperatures have risen by 1.8°F.
- Sea-level rise has increased by an annual rate of 3.2 mm since 1993.
- Glaciers have decreased by more than 60 feet since 1980.
- Sea ice in the Arctic has shrunk by 40% since 1970, and carbon dioxide emissions have risen by 25% since 1958 (National Oceanic and Atmospheric Administration, 2022).

Because of these climate changes (CCs), the United States has experienced 24 droughts, 46 hurricanes, 30 floods, 128 severe storms, and 16 wildfires, all of which have caused billions of dollars in losses. It is also the case that 97% of climate scientists have documented that CC is driven by human activities (Goldberg et al., 2019).

Given increasing evidence of these CC effects on people's lives, Americans who are "very" or "extremely" sure that CC is happening now are about 8 to 1 (57% versus 7% of the population) (Leiserowitz et al., 2021a). More than half of Americans (55%) think that people in the United States are being harmed by global warming "right now," and about half (52%) say that they have experienced the effects of global warming.

Young People's Concern about Their Future World

Young people are increasingly aware that CC impacts are now and will adversely affect their lives in the future, leading them to perceive the need to

DOI: 10.4324/9781003335276-1

take action to demand changes in status-quo energy, transportation, agriculture, building design, and political/legal systems. A survey of young people aged 16–25 in ten counties found that 59% were extremely or very worried and 84% were moderately worried, resulting in 45% experiencing anxiety about the effects of CC on their lives (Hickman et al., 2021). Their experience of eco-anxiety derived from their belief that older generations, institutions, and governments have failed to address CC, particularly impacts on poorer people/countries and people of color (Hickman et al., 2021; Jones & Davison, 2021).

For example, in an interview, Shiva Rajbhandari (2022) described how when he was in seventh grade in 2017, he took an earth science class in his Boise, Idaho school where he learned about CC. Then in 2019, he attended a Global Climate Strike with 1,500 people in Boise, leading to participating in the Sunrise Movement that included protests at a Chase bank demanding they divest in fossil fuel companies. This led to his working on a webcast, Earth Day Live: Boise.

In 2021, he and his peers took a course at Boise State University on CC, given that Jane Fonda agreed to pay their course tuition fees if they would contact their congressional representative regarding the need for Idaho to divest from fossil fuels. For the final projects in the course, students created an online platform for sharing climate stories and an art show. Shiva's actions are driven by his sense that "we can decide the whole fate of the world if we act" (p. 33).

Shiva and other young people increasingly recognize the value of using media to communicate the need for actions to address the climate crisis. Our primary reason for editing this book is that we believe it is important to focus on how youth are engaged in responding to and producing media, given the need to adopt meaningful sustainability practices relatively soon.

Youth are also increasingly interacting with media produced by adults and youth, serving as role models for their productions. For example, they are inspired by youth climate activists such as Greta Thunberg, who, at the age of 15, began a School Strike for Climate in 2018, outside the Swedish parliament. She then documented the strike on Instagram, which then went viral, leading to the launch of the international Fridays for Future strikes *fridaysforfuture.org* from school on Fridays (Hawley, 2022). In addition, during the COVID-19 pandemic, youth increasingly turned to YouTube, TikTok, Instagram, and Twitter to voice their CC messages (Sorce & Dumitrica, 2021). These social media platforms foster engaging interactions between users, contributing to engaging in collective actions, for example, participating in protests.

By adopting these activist stances, youth may challenge status-quo norms and practices leading to changes in institutional practices for addressing the climate crisis. As Beckerman notes:

> Young people need to know how to employ these tools to make a change: For the vanguards of the present dreaming up new ways to fight global warming… this is an essential point: that the shape and extent of the change they seek depend as much on the tools they use as it does on their own will and hunger.
>
> (*Beckerman, 2022, p. 2*)

At the same time, young people are critical of the lack of media coverage of CC, as reflected in the finding that only 57% of Americans indicate that they experience coverage of CC in the media at least once a month (Leiserowitz et al., 2021b). They are also concerned about how social media platforms, corporations, news organizations, and the government spread misinformation about CC (Leiserowitz et al., 2021b).

Fostering Youth Media Production

These concerns about CC have led many young people to assume activist stances on the need to address the climate crisis through digital media—videos, social media, blogs, podcasts, and digital art/music, given the centrality of digital media for interacting with each other (Beach & Smith, 2020; Rousell et al., 2017; Rousell et al., 2022). As of 2021, 13- to 18-year-olds engage in about an average of eight-and-a-half hours of screen media daily, with 8- to 12-year-olds devoting five-and-a-half hours daily (Common Sense Media, 2022). Six in ten view online videos daily, with YouTube, Snapchat, and TikTok being the most popular platforms. In addition, Gen Z people (born after 1996) are more likely to communicate online about CC and assume more activist stances than members of older generations (Tyson et al., 2021). For example, 56% of Gen Z frequently access references on social media about CC; 69% of Gen Z social media users experienced being anxious about the future based on social media messages; and 45% have engaged in social media communication about CC (Tyson et al., 2021).

Youth employ digital media to portray images of flooding, hurricane/severe weather damage, droughts, and forest fires (Beach, 2015; Dunaway, 2015; Newell et al., 2016). Effective media production involves the use of visual portrayals of non-human material objects or events, for example, portrayals of droughts impacting agriculture (Thevenin, 2020).

Young people also know that they can connect with a larger local and or global audience using media as opposed to traditional print media. They can then use media to inform audiences about the need to address local CC effects in their communities (Beach & Smith, 2020; Beach et al., 2017), for example, by producing videos such as The Earth is Melting! (Young People's Trust for the Environment) *youtube.com/watch?v=tlKS9L507VM* and Hurricane Damage and Climate Strikes in Florida (Youth & Climate Action for the Climate Emergency) *youtube.com/watch?v=ODX-izAg1Rg&t=3s.*

As reported in Steve Goodman's Chapter 10, students in the Education Video Center created documentaries about environmental justice issues that adversely impact low-income neighborhoods such as those portrayed in *Shame on You! That Can Be Reused! vimeo.com/369988483* (Goodman, 2020).

We also recognize the importance and value of youth writing as media for communicating with their audiences (Share, in press), as did young writers Jean Hinchliffe (2021) and Jamie Margolin (2020) (Hawley, 2022). As illustrated in several chapters of this book, students may "engage in ecowriting through reflective journaling on environmental themes, work with natural materials in creative ways, or use contemporary media to raise awareness via myth-busting or counter-narrative storytelling" (López et al., 2022, p. 469).

Teachers Supporting Students' Use of Media

As youth voice increased concern about CC effects, teachers can support students by having them employ media to assume activist stances and communicate with authentic audiences about their environmental concerns. Teachers can help students draw on scientific information about CC effects to produce valid information about CC (Leal Filho et al., 2019).

They can also provide instruction on production techniques, for example, how to employ visual images or graphics to achieve rhetorical uptake, for example, portraying how sea rise is related to melting Arctic/Antarctica glaciers caused by higher temperatures. As Walsh (2015) notes:

> Climate visuals are never displayed in a vacuum; they are always displayed by particular groups in service of particular political goals—whether those goals are as focused as securing the reputation and funding of a climate research center, or as broad as rallying transnational opposition to the Kyoto and Lima accords.
>
> *(p. 364)*

Teachers can also help students connect portrayals of those images with people and institutions assuming the need to take action to address the climate crisis, for example, by portraying people in their school or community engaged in promoting the use of clean energy options.

Teachers also engage students in adopting critical perspectives on how narratives shape their perspectives on the natural world. For example, they may examine and critique how the narrative of conquering the West in America led to the degradation of the natural world (Stibbe, 2020) so that "students to explore how our language, stories, metaphors, and representations shape our relationships with the natural world through the power of ideology" (López et al., 2022, p. 466). Students may then create media portraying futurist thinking to imagine and act on alternative perspectives to reconstruct systems, as evident in cli-fi literature (Young, 2022).

Students also need to adopt critical stances on status-quo energy, transportation, agriculture, urban/building design, and political/legal systems which lead to students reimagining and redesigning these systems as well as ecosystems (Wapner & Elver, 2016) through "ecological imagination" (Fesmire, 2010), given how digital tools portray or simulate ecosystems (Hörl, 2017). For example, the EcoMUVE curriculum project *t.ly/b1D7G* provides students with virtual, visual experiences in ponds or forests as ecosystems to analyze changes in these systems *t.ly/Avm2G* (Dede et al., 2021; Smith et al., 2021). Digital tools also immerse audiences in the destruction of ecosystems, as evident in time-lapse video recordings of the ice melts in glacial sites *extremeicesurvey.org/about-eis* or experiences with people coping with CC as portrayed in the Climate Visuals project *climatevisuals.org*. As described in Linda Buturian's Chapter 5, her students create digital stories *tinyurl.com/wqmgz9j* about CC impacts on local rivers and lakes that serve to immerse their audience in the experience of these impacts (Buturian, 2016).

Another challenge for integrating media production within the larger media literacy context is that much of media literacy education does not focus on CC (Leal Filho et al., 2019). Students must acquire critical media literacy perspectives to know how to employ their media to critique status-quo systems. For example, knowing how to critique misinformation on social media regarding CC so that they generate social media that counters this misinformation. They also need to acquire criteria for critically assessing media production, for example, the use of cinematic techniques in video production, to apply those criteria to their video or digital storytelling productions (Smith, 2019).

It is also essential that teachers provide students with multiple options for media production based on, for example, their extensive use of TikTok, Instagram, or YouTube videos as well as their use of social media for interacting with their peers (Bernier, 2020). This book, therefore, includes chapters from various media for fostering the use of different media—video, digital storytelling, social media, art, music, and writing.

Teachers can also support students collaboratively working with people from different backgrounds to create digital media. For example, students from Cincinnati, Ohio, and the Navajo Nation in Arizona shared their experiences related to Native American sustainability practices to create a podcast, Our Future on Fire *t.ly/yt7X* (Bernier, 2020).

Purpose and Organization of This Book

The overall purpose of this book is to provide teachers with examples of youth producing media about CC leading to readers supporting youth in schools and organizations to produce their media. We have organized this book to appeal to an educator audience to provide educators with a sense of purpose by determining how students acquire certain literacy practices for engaging students in media production.

The book focuses on three sections:

1 *Justification for engaging students in media production* related to the need to promote and justify the value of producing media for reaching both local and global audiences. Given how teachers often need to cover a range of different topics related to state and local standards, there is a need to provide teachers with justifications for devoting classroom time to media production about CC. The chapters in this section document the benefits of providing students with opportunities to produce media with specific examples of instruction or projects leading to students generating media. The chapters also describe how youth are motivated to engage in media production, given their concern about the need to address local and global issues of environmental justice.

2 *Critical thinking/learning fostered through media production* about the need to address the climate crisis. Chapters in the second section describe how students learn to engage in critical thinking through media production experience. Critical thinking includes knowing how to engage in systems thinking related to critiquing the impacts of the energy, transportation, agriculture, urban design, political, and legal systems on ecological systems. Through their media production, students are learning new ways of critically thinking about CC and using visual images to convey their messages about CC. They also learn to adopt different roles in productions, such as writer, producer, director, cameraperson, or editor for video/digital storytelling productions (Jiang et al., 2019).

3 *Methods for engaging students in media production* through specific production and writing practices. Chapters in the third section provide teachers with specific methods and examples of media production related to the use of cinematic, social media, artistic, and music techniques related to achieving audience uptake, as well as methods for planning their media productions through brainstorming about the purpose and focus of their productions as well as writing scripts or creating storyboards. For example, for creating documentaries about people engaged in addressing CC, students need to acquire strategies for editing footage of images and interviews with people.

We have also created the website *youthmediaclimatecrisis.pbworks.com* organized according to the three sections that provide readers with links to additional resources, activities, and readings, as well as summaries of each chapter and, in some cases, authors' additional content for their chapters. Readers can also find other links, activities, and reading on a website for Richard's co-authored book (Beach et al., 2017): *climatechangeela.pbworks.com.*

References

Beach, R. (2015). Imagining a future for the planet through literature, writing, images, and drama. *Journal of Adolescent & Adult Literacy, 59*(1), 7–13.

Beach, R., Share, J., & Webb, A. (2017). *Teaching climate change to adolescents: Reading, writing, and making a difference.* Routledge.

Beach, R., & Smith, B. E. (2020). Using digital tools for studying about and addressing climate change. In P.M. Sullivan, J.L. Lantz, & B.A. Sullivan (Eds.), *Handbook of research on integrating digital technology with literacy pedagogies* (pp. 346–370). IGI Global.

Beckerman, C. (2022, February 13). Radical ideas need quiet spaces. *The New York Times Review*, p. 2.

Bernier, A. (2020). Wanting to share: How integration of digital media literacy supports student participatory culture in 21st century sustainability education. *Journal of Sustainability Education, 24.* ISSN: 2151-7452.

Buturian, L. (2016). *The changing story: Digital stories that participate in transforming teaching & learning.* University of Minnesota Libraries Publishing. https://tinyurl.com/sg56r4o

Common Sense Media. (2022). The Common Sense Census: Media use by tweens and teens, 2021. Author. https://www.commonsensemedia.org/research/the-common-sense-census-media-use-by-tweens-and-teens-2021

Dede, C., Grotzer, T., Kamarainen, A., & Metcalf, S. (2021). Virtual reality as an immersive medium for authentic simulations: The case of EcoMUVE. In D. Liu, C. Dede, R. Huang, & J. Richards (Eds.), *Virtual reality, augmented reality, and mixed reality in education.* Springer.

Dunaway, F. (2015). *Seeing green: The use and abuse of American environmental images.* University of Chicago Press.

Fesmire, S. (2010). Ecological imagination. *Environmental Ethics, 32*(2), 183–203. doi: 10.5840/enviroethics201032219

Goldberg, M. H., van der Linden, S., Ballew, M. T., Rosenthal, S. A., Gustafson, A., & Leiserowitz, A. (2019). The experience of consensus: Video as an effective medium to communicate scientific agreement on climate change. *Science Communication, 41,* 659–673. doi:10.1177/1075547019874361

Goodman, S. (2020, April). Teaching for environmental justice at the Educational Video Center. *Journal of Sustainability Education.* http://t.ly/JtT-

Hawley, E. (2022). *Environmental communication for children: Media, young audiences, and the more-then-human world.* Springer.

Hickman, C., Marks, E., Pihkala, P., Clayton, S., Lewandowski, R. E., Mayall, E. E., Wray, B., Mellor, C., & van Susteren, L. (2021). Climate anxiety in children and young people and their beliefs about government responses to climate change: A global survey. *Lancet Planet Health, 5,* e863–e873.

Hinchliffe, J. (2021). *Lead the way: How to change the world, from a teen activist and school striker.* Pantera Press.

Hörl, E. (2017). Introduction to general ecology: The ecologisation of thinking. In Hörl, E., & Burton, J. E. (Eds.), *General ecology. The new ecological paradigm* (pp. 1–74). Bloomsbury.

Jiang, S., Smith, B. E., & Shen, J. (2019). Examining how different modes mediate adolescents' interactions during their collaborative multimodal composing processes. *Interactive Learning Environments, 29*(5), 807–820. doi:10.1080/10494820.2019.1612450

Jones, C. A., & Davison, A. (2021). Disempowering emotions: The role of educational experiences in social responses to climate change. *Geoforum, 118*, 190–200.

Leal Filho, L., Lackner, B., & McGhie, H. (Eds.) (2019). *Addressing the challenges in communicating climate change across various audiences.* Springer.

Leiserowitz, A., Maibach, E., Rosenthal, S., Kotcher, J., Neyens, L., Carman, J., Marlon, J., Lacroix, K., & Goldberg, M. (2021a, September). Consumer activism on global warming. Yale Program on Climate Change Communication. http://t.ly/OngK

Leiserowitz, A., Maibach, E., Rosenthal, S., Kotcher, J., Neyens, L., & McGhie, M. (Eds.) (2021b). *Addressing the challenges in communicating climate change across various audiences* (pp. 1–11). Springer.

López, A., Redmond, T., & Share, J. (2022). Ecomedia literacy, ecojustice, and media education in a postpandemic world. In Y. Friesem, U. Raman, I. Kanižaj, & G. Y. Choi (Eds.), *The Routledge handbook of media education futures post-pandemic* (pp. 463–470). Routledge. doi:10.4324/9781003283737

Margolin, J. (2020). *Youth to power: Your voice and how to use it.* Hachette Books.

National Oceanic and Atmospheric Administration. (2022). Climate change impacts. https://www.noaa.gov/education/resource-collections/climate/climate-change-impacts

Newell, R., Dale, A., & Winters, C. (2016). A picture is worth a thousand data points: Exploring visualizations as tools for connecting the public to climate change research. *Cogent Social Sciences, 2*(1), 1201885. doi:10.1080/23311886.2016.1201885

Rajbhandari, S. (2022). Textbook example. *Sierra, 107*(3), 33.

Rousell, D., Cutter-Mackenzie, A., & Foster, J. (2017). Children of an Earth to come: Speculative fiction, geophilosophy and climate change education research. *Educational Studies, 53*(6), 654–669.

Rousell, D., Wijesinghe, T., Cutter-Mackenzie-Knowles, A., & Osborn, M. (2021). Digital media, political affect, and a youth to come: Rethinking climate change education through Deleuzian dramatisation. *Educational Review, 75*(1), 33–53.

Share, J. (Ed.) (in press). *For the love of writing: Ecowriting in every classroom.* Peter Lang.

Smith, B. E. (2019). Mediational modalities: Adolescents collaboratively interpreting literature through digital multimodal composing. *Research in the Teaching of English, 53*(3), 197–222.

Smith, B., Beach, R., & Shen, J. (2021, August). Fostering student activism about the climate crisis through digital multimodal narratives. *Journal of Sustainability Education.* http://t.ly/QdbJ

Sorce, G., & Dumitrica, D. (2021). #fighteverycrisis: Pandemic shifts in Fridays for Future's protest communication frames. *Environmental Communication.* doi:10.1080/17524032.2021.1948435

Stibbe, A. (2020). *Ecolinguistics: Language, ecology, and the stories we live by.* Routledge.

Thevenin, B. (2020). Engaging with things: Speculative realism and ecomedia literacy education. *Journal of Sustainability Education, 23.* ISSN: 2151-7452.

Tyson, A., Kennedy, B., & Funk, C. (2021). Gen Z, millennials stand out for climate change activism, social media engagement with issue. Pew Research Center. https://www.pewresearch.org/science/2021/05/26/gen-z-millennials-stand-out-for-climate-change-activism-social-media-engagement-with-issue/

Walsh, L. (2015). The visual rhetoric of climate change. *WIREs Climate Change, 6,* 361–368. doi:10.1002/wcc.342

Wapner, P., & Elver, H. (Eds.) (2016). *Reimagining climate change.* Routledge.

Young, R. L. (Ed.) (2022). *Literature as a lens for climate change: Using narratives to prepare the next generation.* Lexington Books.

SECTION I

Justifying the Value of Media Production to Address the Climate Crisis

Educators across the disciplines are concerned about the need and value of teaching about climate change in their classes. At the same time, teachers often do not focus on teaching about climate change, as a national survey found that teachers only devoted approximately 1.5 class hours a year, primarily in earth science classes (Plutzer et al., 2016).

Given the need to adhere to their state and district standards, teachers may need some justification for teaching about climate change for the relatively time-consuming focus on having students produce media within their curriculum, given expectations for covering various topics and subjects in their courses. The Next Generation Science Standards (NGSS) include many standards related to climate change *t.ly/7Lbe*. Moreover, 24 states, or 35% of public schools, have generated standards drawing on the NGSS, but national and state standards for English language arts, social studies, math, art, and music may not include standards related to climate change.

The National Council of Teachers of English (2019) did pass a resolution on the need to "lead students to engage thoughtfully with texts focusing on social and political debates surrounding climate change, and work with teachers in other fields to implement interdisciplinary instruction on climate change and sustainability" (NCTE, 2019, np). Furthermore, the National Council of Social Studies (2019) posits the need for instructions on climate change, given the importance of infusing social justice, historical, and cultural perspectives in teaching climate change.

DOI: 10.4324/9781003335276-2

Justifying the Value of Students Media Production

Teachers across the curriculum need to justify the pedagogical value for engaging students in media production about climate change. This book's introductory section includes chapters that provide this justification by positing the value of youth engaging in producing media across all subjects related to the need to address the climate crisis.

As noted in our book's introduction, one primary justification is that students now use media extensively, given that adolescents average about seven-and-a-half hours daily on their screens (Rideout & Robb, 2019). In the year prior to the pandemic, 89% of students indicated that they used digital tools few days a week, and 42% wanted to use them more often, with 85% of teachers and 96% of principals supporting the increased use of digital tools (Calderon & Carlson, 2019).

Teachers can justify the value of devoting time to producing media by engaging students in responding to media to prepare them for producing media to enhance their confidence in communicating with audiences to take action to address the climate crisis (Cameron et al., 2021). Viewing videos about climate change can lead to an increased environmental concern, resulting in positive shifts in viewers' attitudes regarding the severity of climate change and the need for action (Janpol & Dilts, 2016). Watching videos about climate change also provides students with visual experiences of impacts as well as scientific data regarding the validity of human-caused climate change (Davis et al., 2020).

Through producing media, students acquire multimodal composing skills for employing argumentative literacy practices, enhancing students' sense of agency, self-efficacy, and attitudes toward enacting change (Littrell et al., 2019, 2020; Rooney-Varga et al., 2014). For example, students used media to portray the value of experimenting with alternative energy sources and strategies for lobbying local governments (Beach & Smith, 2021). Students engaged in a program in Puerto Rico to produce videos based on issues of plastics pollution, coastal erosion, and sustainable agriculture, leading to their perceiving need for themselves to take action to address these impacts (Leckey et al., 2021). Students also created videos on local sustainability issues, including a video, Green Grease Guzzlers *t.ly/io_V*, that documents the adverse impacts of using gas for transportation leading to using alternative fuels for use with a food truck (Corwin, 2020). One student from this project noted how the activity of creating videos "made me feel like I was a part of changing it. I was doing something about it instead of just saying, hey, we should do something about it" (np).

Students also benefit from participation in school or organization projects, providing them with training and support (Littrell et al., 2019, 2020; Tayne et al., 2020). To create these projects, teachers may draw on organizations such as Action for the Climate Emergency acespace.org, Young People's

Trust for the Environment *ypte.org.uk*, The UK Youth Climate Coalition *ukycc.com*, Climate change Education *earth.stanford.edu/climate-change-ed*, Youth4Climate *youth4climate.info*, Young Voices for the Planet *youngvoices-fortheplanet.com*, Our Climate Our Futures *ourclimate.us*, Connect4Climate *connect4-climate.org*, and the Climate Reality Project *climaterealityproject.org*

Justifying the value of youth media production entails documenting what students learn *through* their participation in media production. For example, an analysis of students producing videos on climate change as part of the Lens on Climate Change (LOCC) project at the University of Colorado identified how by producing their videos, students engaged in reframing their conceptions of climate change based on researching topics portrayed in their videos (Littrell et al., 2020, 2022). One student noted that she had a "stronger belief that the water in [my home state] will run out instead of returning" (Littrell et al., 2020, p. 602).

In the LOCC program, the students generate their narratives through film, often sharing with their community the problem and potential solutions (Littrell et al., 2020, p. 596). For example, four Native American students produced a video, "Coyote and the Drought," about the lack of water in their hometown of Gallup, New Mexico. They created drawings and video clips of related scenes representing their home community (Littrell et al., 2020). The students decided that, given their audience as children, to frame their video on a Navajo story that distinguishes humans' excessive use of water from how a coyote employs minimal use of water:

> Students in the program "left the program feeling inspired to continue learning and/or communicating with others about climate change or inspired to engage in actions to help mitigate climate change or other environmental challenges in their communities."
>
> (*Littrell et al., 2022, p. 3*)

In producing media, students are also learning to reimagine systems by using media to *reframe* their status-quo perceptions of environmental impacts related to the need to assume more activist stances to communicate with audiences beyond the classroom (Littrell et al., 2022). In one project, high school students created videos portraying our dependency on automobiles in transportation systems leading to reimagining a transportation system based on enhanced public transportation and carpooling (Karahan & Roehrig, 2014). Similarly, students in Allen Webb's courses at Western Michigan University create websites, blogs, wikis, and cli-fi novels, as well as online "climate manifestos" positing the need for action to be read at college or community protests (Beach et al., 2017; Polk et al., 2023).

To produce media reframing status-quo systems, students may draw on responses to media or cli-fi literature portraying transitions to alternative, more sustainable futures, as portrayed in *The Ministry for the Future* (Robinson, 2020), in which characters enact change as opposed to literature that portrays only disaster scenarios (Mackenthun, 2021). For example, in the novel set between 2030 and 2048, the main character, as the President of the Ministry for the Future, creates a new currency system based on the degree to which products lower CO_2 emissions resulting in the production of ecological products and reforestation, leading to emissions reduction.

Through their media production, students also experience an enhanced *re-seeing* of environmental effects, given their need to create visual portrayals of climate change effects in their videos. Engaging in these practices in the LOCC project led to their "*re-enactment* by shaping their thinking around how they could help their communities either through direct action or continued communication with family or community members about climate change and other environmental challenges after the program" (Littrell et al., 2022, p. 18), as evident in these youth-produced videos:

- Your Voice, from the Climate Change Initiative, University of Massachusetts, Lowell *vimeo.com/70140870*
- Frontline Youth: Fighting for Climate Justice Climate Justice Alliance *vimeo.com/387928607*
- Connect4Climate meeting *youtube.com/watch?v=iHfCJybzw9k*

These youth-produced videos reflect students' ability to generate engaging videos for educating their audiences about the negative impacts of climate change implying audiences' need to take action. Students' participation in the LOCC project and other similar projects and in their classrooms suggests the value of using media to foster students' willingness to address the climate crisis. The chapters in this section amplify these justifications by providing examples of students acquiring valued practices through media production for learning to address the climate crisis.

Chapter Summaries

The first chapter in this section, "We Are Nature Defending Itself: Universal Climate Literacy DIY with Youth Media Productions and Engagement" by Marek Oziewicz and Scott Spicer, posits the need to address climate change, given older generations' alignment with the consumer economy. They recommend transforming instruction to foster youth media production to generate a counter-narrative to this status-quo economic system through innovative instructions in climate literacy practices. The chapter justifies the benefits and value of employing these climate literacy practices for engaging students in media-production projects to foster their self-efficacy related to becoming activists in addressing the climate crisis.

The second chapter, "Centering Utopia: Fostering Youth Climate Change Education by Exploring and Envisioning Hopeful Futures" by Michelle Jordan, Catherine Lockmiller, and Steven J. Zuiker, formulates a justification for engaging students in media production to communicate their concern about the need for engaging in utopian thinking for addressing the climate crisis through the use of digital and literacy media for portraying and imagining adaptation and mitigation practices. In a summer solar energy engineering program, students acquired climate literacy practices based on renewable energy systems' transitions, and information about the use of video games, film, music, comics, books, podcasts, and other media related to climate change.

The third chapter, "General Ecology and Speculative Pedagogies: Youth Digital Media Practices for Climate Justice" by David Rousell, Thilinika Wijesinghe, and Amy Cutter-Mackenzie-Knowles, describes the use of media for engaging in systems thinking for active participation in addressing issues of environmental justice or participating in international protest events. The authors describe the use of the Climate Action Adventure! web application at their university for connecting virtual learning environments with live social media platforms and climate data for fostering the practices of anonymous participation, character-building, and world-building. They also pose a curriculum framework for addressing climate justice issues related to students creating and sharing digital texts with audiences.

The final chapter in this section, "Integrating Community-Based Participatory Research Approaches with Climate Justice Digital Media Projects" by Emily Polk, describes examples of students in her Intro to Environmental Justice composition class at Stanford University creating digital media—podcasts, blogs, digital images/maps, etc., based on their investigations of local climate justice issues, for example, reporting on issues of water governance from an indigenous perspective. In addition, students in her course engaged in case-study projects studying local Bay Area sustainability issues, leading to students creating podcasts, blog posts, and digital presentations for communicating with their peers and media publications about members of the communities they studied to achieve uptake from local and global audiences.

For additional links, resources, and readings related to this section, please refer to the book's website *t.ly/TUmH*

References

Beach, R., Share, J., & Webb, A. (2017). *Teaching climate change to adolescents: Reading, writing, and making a difference.* Routledge.

Beach, R., & Smith, B. (2021). Analysis of online videos by and about adolescents addressing climate change. Paper presented at the Annual Meeting of the Literacy Research Association, Atlanta, Georgia.

Calderon, V. J., & Carlson, M. (2019, September 12). Educators agree on the value of ed tech. Gallup. Retrieved from https://www.gallup.com/education/266564/educators-agree-value-tech.aspx

Cameron, L., Rocque, R., Penner, K., & Mauro, I. (2021). Evidence-based communication on climate change and health: Testing videos, text, and maps on climate change and Lyme disease in Manitoba, Canada. *PLoS ONE, 16*(6), e0252952. doi:10.1371/journal.pone.0252952

Corwin, J. (2020, May). Green guerrillas youth media tech collective: Sustainable storytellers challenging the status quo. *Journal of Sustainability Education, 23.* http://t.ly/9VW1

Davis, L. S., León, B., Bourk, M. J., & Finkler, W. (2020). Transformation of the media landscape: Infotainment versus expository narrations for communicating science in online videos. *Public Understanding of Science, 29*(7), 688–701. doi:10.1177/0963662520945136

Janpol, H. L., & Dilts, R. (2016). Does viewing documentary films affect environmental perceptions and behaviors? *Applied Environmental Education & Communication, 15*(1), 90–98. doi:10.1080/1533015X.2016.1142197

Karahan, E., & Roehrig, G. (2014). Constructing media artifacts in a social constructivist environment to enhance students' environmental awareness and activism. *Journal of Science Education and Technology, 24*(1), 103–118.

Leckey, E. H., Littrell, M. K., Okochi, C., González-Bascó, I., Gold, A., & Rosales-Collins, S. (2021). Exploring local environmental change through filmmaking: The Lentes en Cambio Climático program. *The Journal of Environmental Education, 52*(4), 207–222, doi:10.1080/00958964.2021.1949570

Littrell, M. K., Gold, A. U., Koskey, K. L. K., May, T. A., Leckey, E., & Okochi, C. (2022). Transformative experience in an informal science learning program about climate change. *Journal of Research on Science Teaching,* 1–25. doi:10.1002/tea.21750

Littrell, M. K., Okochi, C., Gold, A. U., Leckey, E., Tayne, K., …, Wise, S. (2019). Exploring students' engagement with place-based environmental challenges through filmmaking: A case study from the Lens on climate change program. *Journal of Geoscience Education 68*(1), 80–93. doi:10.1080/10899995.2019.1633510

Littrell, M. K., Tayne, K., Okochi, C., Leckey, E., Gold, A. U., & Lynds, S. (2020). Student perspectives on climate change through place-based filmmaking. *Environmental Education Research, 26*(4), 594–610. doi:10.1080/13504622.2020.1736516

Mackenthun, G. (2021). Sustainable stories: Managing climate change with literature. *Sustainability, 13,* 4049. doi:10.3390/su13074049

National Council of Teachers of English. (2019). Resolution on literacy teaching on climate change. Author. Retrieved from http://t.ly/VFMu

National Council of Teaching Social Studies. (2019). Teaching climate change. Author

Plutzer, E., Hannah, A. L., Rosenau, J., Mclimate changeaffrey, M. S., Berbeco, M., & Reid, A. H. (2016). *Mixed messages: How climate is taught in America's schools.* National Center for Science Education. Retrieved from http://ncse.com/files/MixedMessages.pdf

Polk, E., Beach, R., & Webb, A. (2023). Global literacies for addressing the climate crisis. In S. N. Kerkhoff & H. A. Spires (Eds.), *Critical perspectives on global literacies: Bridging research and practice* (pp. 230–246). Routledge.

Rideout, V., & Robb, M. B. (2019). The Common Sense census: Media use by tweens and teens, 2019. Common Sense Media. Retrieved from https://tinyurl.com/y3thz562

Robinson, K. S. (2020). *The ministry for the future.* Hachette.

Rooney-Varga, J. N., Brisk, A. A., Adams, E., Shuldman, M., & Rath, K. (2014). Student media production to meet challenges in climate change science education. *Journal of Geoscience Education, 62*(4), 598–608. doi:10.5408/13-050

Tayne, K., Littrell, M. K., Okochi, C., Gold, A. U., & Leckey, E. (2020). Framing action in a youth climate change filmmaking program: Hope, agency, and action across scales. *Environmental Education Research*, 1–21. doi:10.1080/13504622.2020.1821870

1

WE ARE NATURE DEFENDING ITSELF

Universal Climate Literacy DIY with Youth Media Productions and Engagement ·

Marek Oziewicz and Scott Spicer

Climate Literacy and the Four Premises

This chapter builds on the assumption that one key goal of today's education—across all subject areas and grade levels—should be to empower young people with the conceptual tools they need to grasp the human-planetary predicament of the Anthropocene. This suggests articulating this understanding as stories that engage anticipatory imagination and mobilize action. We propose that this specific set of conceptual tools is captured by the notion of "climate literacy": a capacity that includes an understanding of facts and numbers (i.e., climate science literacy), but centers on developing values, attitudes, and behavioral changes necessary to build sustainable futures (Oziewicz, 2022b).

The argument rests on four premises. One, that Earth is in a state of human-caused climate emergency whose scope, speed, feedback loops, and impacts are so complex and so "massively distributed in time and space relative to humans" (Morton, 2013, p. 1) that they are almost incomprehensible to the vast majority of people alive today. Two, that the main systems of our global civilization—its legal, economic, political, educational, technological, and social structures—were designed without concern for the biosphere and designed to exploit it indefinitely. These systems have proven to be incapable of redesigning themselves to the extent that can slow down, let alone reverse, the accelerating demolition of the planet's life support systems. Three, that human cognitive architecture is evolved for narrative understanding, which means that climate change and other urgencies of the Anthropocene are primarily challenges to our story systems rather than challenges to our technology, politics, or economy. And four, that a transition to an ecological civilization depends on our ability to help

DOI: 10.4324/9781003335276-3

young people exercise their voice and agency in the present so that they are able to create alternatives to the dominant, hegemonic narrative that legitimizes ecocide. This is especially the case in spaces like digital media productions, which young people navigate with uncanny expertise. Given that the self-serving illusions of the petronormative order are maintained by the dominant rhetoric of eternal growth in the market economy, defusing and creating alternatives to this dominant narrative is a prerequisite for any meaningful change.

Ecomedia Literacy and Youth Digital Productions

Extrapolating from the four premises, we believe that our world's future will, in no small way, be shaped by young people's ability to use media to address the climate emergency. The term for this emerging skill set is "ecomedia literacy." As explained by Antonio López, one of the leading scholars in the field (see also Chapter 9 in this book), ecomedia literacy "conceptualizes media as an ecologically embedded ecosystem" (López, 2021, p. xv) with its own ecological footprints and mindprints, thus grounding media production and consumption, at once, in "the physical environment inhabited by humans and nonhuman alike" (p. 10) and in the socio-cultural and economic environments of media users (p. 11).

Ecomedia literacy decolonizes traditional media studies, offering an activist alternative to traditional media literacy and education, which are rooted in "a way of thinking about and being in the world that is the source of our planetary ecological crisis" (p. xv). Ecomedia-literate young people subvert the status quo by rejecting the fake abstractness of what López calls "medialandia"— "*information* and *entertainment technologies* with no material reality or relationship with the physical world" (p. 11)—in favor of real-world-grounded "ecomediasystems," where the use of media helps them appreciate the eco-ethical relationship between ecomedia objects, environments, and themselves.

This opportunity to learn to see the world "as relationships, interconnections, and systems" (p. 93) within specific ecomediasystems and the larger "ecomediasphere" (p. 124) is the key reason why young people are increasingly drawn to digital media productions. It is with these tools that youth are searching for productive methods of eco-citizenship that lays the foundations for the possibility of ecocentric futures. Unfortunately, traditional channels within the schooling systems have proven inadequate to accommodate student concerns, leading to digital spaces becoming a creative force for young people's DIY climate literacy education.

In digital media, young people's control of the narrative enables them to call out the ongoing ecocide, articulate counter-narratives to it, and imagine the ecocentric future they want to design. That said, climate literacy should not be a DIY project. It should be centered in 21st-century education and promoted as experiential, dialogical do-it-with-others (DIWO) learning. We believe that actively supporting youth media productions and engagement is essential if teachers are serious about helping young people fight for climate justice, universal climate literacy, and a transition to an ecological civilization.

One of the paradoxes of the climate emergency is that those with the most stake in the future have the least say in policy, economics, and other decisions that shape the possibility or impossibility of that future. As a result, over the past 30 years or so, young people have continued to be cheated out of their future. Stealing someone's future may sound like a science fiction concept. However, we find it an accurate description of the process by which neoliberal economy, national and international legal systems, politics, agribusiness, industry, and other systems of the global petrocivilization are actively eroding the possibilities of livable futures for today's children and whoever comes after us.

Young people know it. And their anger is real. For example, over the past three years, I (Marek) have asked students in my courses to answer a few questions about the two opening chapters of Geoff Rodkey's middle-grade novel *We're Not from Here* (2019). One set of questions concerns the students' feelings toward the human refugees who are seeking asylum on the planet Choom. The first time the question is asked, the students are made to understand that the refugees from destroyed Earth are just ordinary humans. In one recent sample of 11 answers (2022), 81.8% of responses indicate sympathy for the refugees (see Figure 1.1).

However, when students are asked to imagine that the refugees are ecological criminals who profiteered on the destruction of Earth, 60% are indifferent or unsympathetic toward these last humans (see Figure 1.2).

When asked, "How do you feel about this group of hyper-privileged people representing all of humanity to alien civilizations? Do you think they represent you?" not a single student felt represented and many were outraged. "I feel anger," wrote one undergrad. "I don't think they represent me. I'd agree

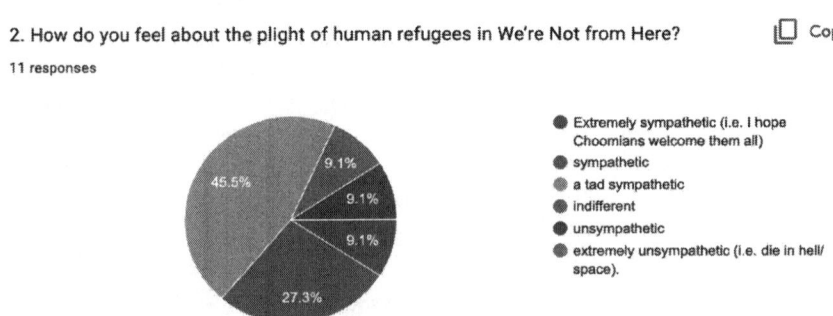

2. How do you feel about the plight of human refugees in We're Not from Here?   Copy

11 responses

- Extremely sympathetic (i.e. I hope Choomians welcome them all)
- sympathetic
- a tad sympathetic
- indifferent
- unsympathetic
- extremely unsympathetic (i.e. die in hell/ space).

45.5% 9.1% 9.1% 9.1% 27.3%

FIGURE 1.1 Students' Perceptions of the Plights of Human Refugees

12. Imagine that Lan and other refugees arriving at Choom are not ordinary humans. Imagine most of them are families of politicians, CEOs, and bankers whose executive decisions led to the ecological collapse of planet Earth. Imagine that other travelers are families of the super rich, from Big Pollute and Wall Street to Narcolords and Oligarchs, whose immense wealth on Earth was built on extractive plunder of the planet and its myriad life forms, including ordinary people. With this knowledge, how do you feel about the human refugees and the case for asylum they make? (NOTE: while children are not responsible for their parents' crimes, in most cases they benefit from inherited wealth and privilege)

10 responses

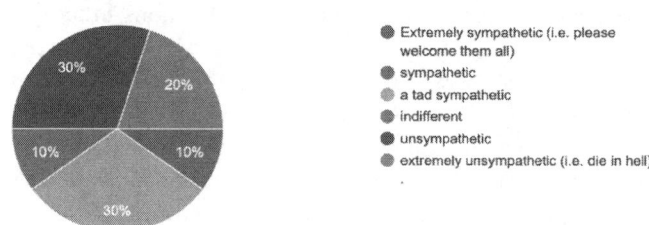

FIGURE 1.2 Students' Perceptions of Human Refugees as Ecological Criminals

with the aliens on not wanting to live with them anymore." Although this thought experiment is limited, it demonstrates the anger young people feel at the thought of being cheated out of a livable future.

Scholars know that this stealing is real. Vast literature exists showing how stealing from the future has become the core operating principle of the neoliberal order and how that order is the key driver of climate change (Bell, 2021; Ghosh, 2016; Haiven, 2014; Joseph, 2017; Klein, 2014; Lent, 2017; Monbiot, 2016; Read & Alexander, 2019; Szeman & Boyer, 2017). We even have a term—"slow violence" (Nixon, 2011, p. 2)—for the attritional wake of delayed destruction this stealing creates for present and future generations. Nor is it a mystery where the notion of stealing from the future originated. Again, vast literature shows how this ideology and practice evolved from a cluster of earlier notions at the heart of the settler-colonial mindset (Ghosh, 2021; Johnson & Wilkinson, 2020; Kimmerer, 2013; Lent, 2021; Shiva & Shiva, 2021; The Red Nation, 2021). This scholarship has also established a clear causal connection between stealing from the future and climate emergencies. In this line of thought, as Amitav Ghosh puts it:

The present phase of the planetary crisis is not new at all. It represents the Earth's response to the globalization of the ecological transformations that were set in motion by the European colonization of much of the world. Those processes have now escaped the boundaries of the three colonized continents and have become planetary forces, in no small part because

Western settler-colonial culture is no longer confined to the settler colonies. Since the adoption in 1989 of the Washington Consensus, the ideologies and practices of settler colonialism have been actively promoted, in their neoliberal guise, by the world's most powerful countries, and have come to be almost universally adopted by national and global elites.

(2021, p. 167)

The big picture about what forces, how, and why have put humanity on a collision course with the Earth's ability to support life is clear. What is less clear is where education fits on the criminal dock among other engines of petronormativity responsible for the ongoing ecocide. Are our education systems as guilty as the neoliberal economic order, the media, or Big Oil? Perhaps not. Has modern education done enough to prepare young people—conceptually, emotionally, socially, and culturally—for the challenges of living in a climate-changed world? Has it empowered them to identify, expose, and resist forces of oppression, exploitation, and corporate-funded denial that legitimize ecocide in the name of profit/progress? Or has it equipped them with the skills to imagine, design, and create an ecological civilization? Our answer is no. We can definitely do better (Oziewicz & Saguisag, 2021; Paulsen et al., 2022; Young, 2022).

Almost 20 years ago, David W. Orr (2004) remarked that "We are still educating the young as if there were no planetary emergency" (p. 27). Disgracefully, this is still the case in 2022. Before we discuss the opportunities this gap creates, we thus need to recognize that our education systems have been systemically complicit in accelerating the climate catastrophe. This complicity is driven by two main operational protocols: ignoring the issue (dis-attention) and reducing climate change to a subfield of science education (ghettoization).

Climate Change and Climate Literacy in Schools

While dis-attention is not denial, teaching about climate change is almost absent in today's curricula. Teachers are not sure if and where climate change fits in their subject areas—except for New Jersey, which, in 2020, became the first state in the United States to mandate climate education at every grade level (Burney, 2020). School boards have no state policies to guide action, and state education departments have no national policies to build on. Leading literacy and educational organizations have acted as if they live on another planet too: a recent study reveals that climate change was ignored at 99.72% of presentations at major literacy conferences during the years 2008–2019 (Panos & Damico, 2021).

Like the vast majority of other countries, the United States has little to no climate education or curricula to speak of. Climate literacy is not adequately covered within environmental education, which we have had since the 1970s (History of Environmental Education, n.d.) or within sustainability education, which we have had since the 1990s (History of Education for Sustainable Development, n.d.). Where (some) learning about climate change is included, such as in the Next Generation Science Standards (since 2013), it is framed as a science-only issue and presented as a (detached, abstract, and limited) form of knowing that no Big Oil executive would ever find threatening: say, "Changes in Earth's tilt and orbit cause climate changes such as Ice Ages" (ESS1.B "Earth and the solar system" for grades 9–12) or "Complex interactions determine local weather patterns and influence climate, including the role of the ocean" (ESS2.D "Weather and climate" for grades 6–8).

This narrow disciplinary focus frames climate change as merely a sub-idea within the Earth Space Science Progression and disconnected from Life Science and Physical Science core ideas. Such framing has the effect of masking the entanglement of climate change with our food systems, legal systems, consumption habits, market practices, dominant ideologies, and other science and non-science spaces of human activity that drive climate change. And while the notion that students in grades 6 and up should have an understanding of how human activities affect global warming is, indeed, mentioned (ESS3.D "Global Climate Change" for grades 6–8), the NGSS vision of this knowledge is, again, limited: (curiously) N/A for students in grades 5 and below, and (merely) an understanding that climate models are imperfect and will continue to be improved for students in grades 9 and up (NGSS, "Appendix E—Progressions Within the Next Generation Science Standards").

No mention is ever made that students might *learn about*, let alone *learn to critique*, the dominant paradigm of carbon-intense perpetual economic growth that is known to be the key driver of climate change. This omission is one testimony to how deeply the NGSS is invested in the technocratic discourse— also used by the dominant petronormative political and economic structures of our world—that is unable to confront the current energy epistemology on its own. Instead of addressing the structural drivers of climate change, it promotes "learning" about climate change in ways that do not challenge the business-as-usual operations of the ecocidal market system as it exists today. Is this really the best we can teach to a generation that grows up in a world of accelerating human-driven climate disasters?

Given the scope of the challenge, it is now increasingly recognized that we need not just *education for climate science literacy* but a deeper, broader, and more interdisciplinary *education in climate literacy*. The year 2017 saw the publication of the first book about climate literacy education: *Teaching Climate Change to Adolescents* by Richard Beach et al. (2017). It also saw the first collection on *Teaching Climate Change in the Humanities*, edited by Stephen Siperstein et al. (2017). In 2019, NCTE became the first professional organization in the field of education to issue a statement on climate change: "Resolution on Literacy Teaching on Climate Change."

In 2020, former education secretaries John King, Arne Duncan, and the Governor of New Jersey, Christine Whitman, issued an open letter to president-elect Joseph Biden, urging him to support climate education nationwide. The letter stressed "the considerable need to address climate change in America's public schools" and called climate education "a critical opportunity toward climate solutions," building resilience, and preparing "children and youth to advance a more sustainable world" (King et al., 2020). And 2021 saw the creation of Climate Lit: a for-teachers and by-teachers online resource hub for teaching climate literacy with children's literature and media across all subject areas and at all grade levels *climatelit.org*. All these efforts are part of the push for building universal climate literacy in America's public education systems. Hopefully, these efforts will lead to a systemic change and transform education in time for it to make a difference. While we appear to be stuck in the suicidal, petronormative race to the bottom, we still have a chance to transition to an ecological civilization.

The other way today's education has so far failed to equip young people with the knowledge they need to adapt to a rapidly changing world is through domain ghettoization of climate change as a STEM issue instead of a worldview issue which it really is (Clayton et al., 2021; Crist, 2013; Ghosh, 2016; Klein, 2021; Lent, 2017). This segregationist, reductionist approach was part of the information-deficit paradigm framework that also drove climate science until quite recently. Within this paradigm, which also informs the Next Generation Science Standards (NGSS, 2013), it was assumed that climate literacy is synonymous with climate science literacy—knowing and being able to communicate the facts. It was also taken for granted that scientific facts are all that governments, corporations, and the public need to know to act.

Unfortunately, the fight over how to respond to climate change is not about scientific rationality. It is about the power of fossil fuel corporations and other industries that actively profit from destroying the planet (Mann, 2021; Supran & Oreskes, 2021). In 2020, for example, global fossil fuel subsidies alone amounted to $5.9 trillion (IMF, 2022). This is almost ten times more than the $632 billion the world spent on global climate action that year (Burg, 2022). In the current system, we are caught up in actively funding the destruction of life on our home planet.

The information-deficit paradigm has been dominant in education too. It was supported by the neoliberal technocratic discourse which locates climate change wholly within the fields of science and economy while ignoring its social, cultural, conceptual, psychological, and other dimensions (Taylor, 2013). Accordingly, 99.88% of all funding available for climate-related research between 1990 and 2018 was directed to the natural and technical sciences. Only .12% of this funding was used to support climate-focused research in social sciences related to the change of attitudes, values, norms, and behaviors: a change that has long been recognized as central to any effective climate policy and action (Overland & Sovacool, 2020).

The difficult thing about climate change is that it is primarily the challenge to our dominant "human-supremacist worldview" that takes the planet for granted and sees all its life forms as subject to human whims (Crist, 2019, p. 3). Given that addressing climate change requires "rapid, far-reaching and un-precedented changes in all aspects of society" (International Panel on Climate Change, 2018, np)—changes that are mostly social, cultural, and political—the humanities and social sciences have a crucial role to play in this process. They are "disciplines that have long attended to the intricacies of social processes, the nature and capacity of political change, and the circulation and organization of symbolic meaning through culture" (Szeman & Boyer, 2017, p. 3).

Specifically, the collective challenge faced by us is to reimagine ourselves as indigenous to nature rather than separate from it, kin to all life rather than its masters. The space for this reimagining is called the story. And it is through and with stories that we need to map out ecocentric ways of being, belonging, and thriving into the future. Doing so can liberate us from suicidal dependence on fossil fuels and other ecocidal processes that have built the global, industrial petrocivilization. We *are* nature. How do we tell this new story? And, where do we start?

Narratives We Live By

The argument for the primacy of stories in shaping human identity, culture, and the course of civilizations is complex but well-established. A vast body of research shows that human cognitive architecture evolved for narrative understanding. "Narrative reflects our mode of understanding events, which appears largely … to be a generally mammalian mode of understanding" (Boyd, 2009, p. 131). We process causality, chronology, memory, events, emotional states, concepts, and all other content of meaning-making processes as stories or components of stories (Herman, 2002, 2013; Hogan, 2003, 2011, 2013; Schank, 1990; Turner, 1996). Our storied understanding frames not only our personal experiences—the so-called "stories of our lives"—but also our grasp of abstract concepts like nationalism (Anderson, 1983; Hogan, 2009), justice (Oziewicz, 2015), worldview (Lent, 2017), economy (Joseph, 2017), political order (Monbiot, 2017), and relationship to the nonhuman (Crist, 2019; Ghosh, 2021; Kimmerer, 2013). It also includes responses to the climate emergency (Klein, 2019; López, 2021; Mann, 2021; Marshall, 2014; Moore, 2016; Norgaard, 2011; Oreskes & Conway, 2010; Oziewicz et al., 2022). Whether called rhetoric, ideology, narrative, secular mythology, legitimizing myth, discourse, or yet something else, story is everything.

Recognizing that "stories-we-live-by" are the primary means by which we navigate reality has three important implications for the argument in this chapter. One—a cornerstone of the environmental humanities—is that the fight for the Earth's future is first and foremost fought in the space of language and imagination. Since most constraints working against climate action are cultural rather than technological, the argument is that we need new ways of thinking, a new ethic of partnership with the nonhuman, and a new story about who we are, as a species, in relation to all other forms of life on the planet.

We must, as a species, find a new, emotionally compelling story "capable of mobilizing social adaptation" to the realities of a climate-altered world (Emmett & Nye, 2017, p. 8). Discovering and articulating this story is our primary task because any social change needs stories to guide its vision of the future. After over 30 years of climate change negotiations, compromises, and toothless agreements, it is clear that politicians or scientists will not introduce real action on climate. Nor is it the kind of change that a group of billionaires can commission. It is the change that can only emerge from the actual practice and conversations among ordinary people, a change that will come from below or not at all.

The second implication is that anyone or any community aspiring to create new stories that seek to bring about a transition to an ecological civilization should be mindful of the dominant stories and aware of the realities these stories help to maintain. A helpful grammar of stories-into-reality can be found in Yuval Noah Harari's *Sapiens*, where he posits that we each inhabit three types of reality: subjective (my beliefs and experiences), intersubjective (beliefs and experiences we share with others), and objective (realities existing irrespective of our awareness, beliefs, or experiences). These three realities mesh with two orders, which Harari calls natural orders (such as the weather) and imagined orders (such as ways to measure temperature in Celsius or Fahrenheit). In this framework, intersubjective imagined orders—Harari's term for complex story and belief systems shared by large populations—are how humans build civilizations and maintain social cohesion (Harari, 2015, pp. 112–118). Nation states, currency, religions, law, economy, politics, corporations, and so much more are all forms of intersubjective imagined orders (pp. 26–32).

Imagined Orders of Denial and Despair

We want to suggest that the two dominant intersubjective imagined orders today are the reality of denial and the reality of despair. Denial is the dominant intersubjective order of our world of politics, business, economy, industry, education, and other main systems of our petrocivilization. This order rests on self-deluded storytelling by individuals and institutions so invested in petronormativity that they refuse to confront the ecocidal status quo even when they appear to take climate change seriously. These stories build on the premise that the current market system is capable of solving the problems of climate change, despite evidence that the same market system is also structurally designed to generate these problems.

Stories of denial include anything from flat denial of climate change to its misattribution to natural causes. These stories make possible subtler forms of legitimizing inaction that include slow-walking change, delay, deflection, greenwashing, and other strategies which promote "a vocabulary of normalized denial" (Zimmerman, 2022). Some of the most successful stories of normalized denial are what Zimmerman calls "Restoration Narratives": stories that promise a miraculous way out in the future through forms of economic growth, technocratic fixes, future technologies, or their combinations.

The other dominant intersubjective imagined order today is despair. Despair presides over our mediasphere: the world of culture, media, films, games, and other forms of collective storytelling. The rise of dystopia—as well as the rise of negative emotions like ecogrief, eco-anxiety, climate fear, solastalgia, withdrawal, and depression—are all products of the intersubjective reality of despair (Oziewicz, 2022a). At the heart of the order of despair is the end-of-times story that we are doomed, that the world of humanity is coming to an end. One collective name for the many stories within this order is "Rapture Ideologies" (Wheal, 2021). According to Jamie Wheal, these ideologies include stories told by ISIS fighters, Christian Zionist, doomsday preppers, and millionaires who invest in New Zealand bunkers.

All of these stories are variations on a single plot anchored in four assumptions: the current world cannot be saved; soon it will collapse; I and mine will make it to the other side; so let us hasten the ending and thus the arrival of the purged world (Wheal, 2021, p. xii). Where stories of denial promise miraculous salvation in the future, stories of despair promote the redemption of the few at the expense of the many. Where the former promise redemption by market forces, the latter expect redemption through divine/supernatural forces. Much of our world's inaction on climate is due to the legitimizing power of restoration narratives and rapture ideologies that have become our dominant intersubjective imagined orders.

Intersubjective Orders of Hope and Youth-Produced Media

What is the alternative? We suggest that today's key challenge is to design an imagined intersubjective order based on activism, vision, and hope I have elsewhere called "planetarianism" (Oziewicz, 2022a). This leads us to the third implication of stories as tools to shape reality: they need to be told by young people, those who have the most at stake in the future. Unlike publishing, formal education, and other spaces dominated by adult gatekeepers, digital media productions offer a venue where young people find the most autonomy, discover their voice, and achieve the most impact.

The need for youth-created (eco)media as a space where young people can share their concerns, perspectives, questions, and insights cannot be overstated. Youth media productions lend themselves to quick sharing that can lead to ad hoc activism or, at a minimum, keep the discussion alive. As López notes in this book (Chapter 9), youth ecomedia productions are a form of experiential learning, "more DIWO (do-it-with-others) than DIY (do-it-yourself)" (p. 195), creating communities of action committed to ecojustice and other values young people are passionate about.

Providing space for students to create their (eco)media narratives in a formalized classroom environment requires pedagogical intentionality. Spicer's (2022) book, *Student-Created Media: Designing Research, Learning, and Skill-Building Experiences*, sets out to aid instructors and support professionals in designing student-created media projects that better engage students in complex topics. Though typically student-created media assignments are designed for students to reinforce subject knowledge acquisition and media literacy skill set development, there are certain instances where the aspirational goal of action becomes more imperative. The issue of climate change, being existential, is just one such case.

This goal is at the apex tier of Spicer's Hierarchy of Student-Created Media Potential (Spicer, 2022a, pp. 8–11) (see Figure 1.3). Though not exclusive, the "Action" tier is where many of the transformative effects of the media project occur. Transformative effects in the context of a media assignment are in essence the enduring impact of the assignment (Spicer, 2022b, pp. 5–7). These are the qualities that last with the students, instructors, support professionals, and external partners long after the project is complete.

Transformative effects can occur on a large or small scale. Examples include:

1 A student takes an internship, volunteership, or job position with an organization due to the relationship they established during the creation of the media project.
2 A student decides to further their education in a certain direction due to the topic they studied while creating their media project.
3 A student decides to change their behavior as a result of what they learned while creating their media project.

Merely the act of creating media itself is not sufficient to move a student from an academic provocation to activism. So how does one design a student-created media assignment as a catalyst for action? We believe that three components are critical. One, it is important that the curriculum must "close the loop." This includes the need for both personal reflection and multi-perspectival curricular opportunities on the students' selected climate change topic. Two, students must do a deep dive on at least one climate change topic.

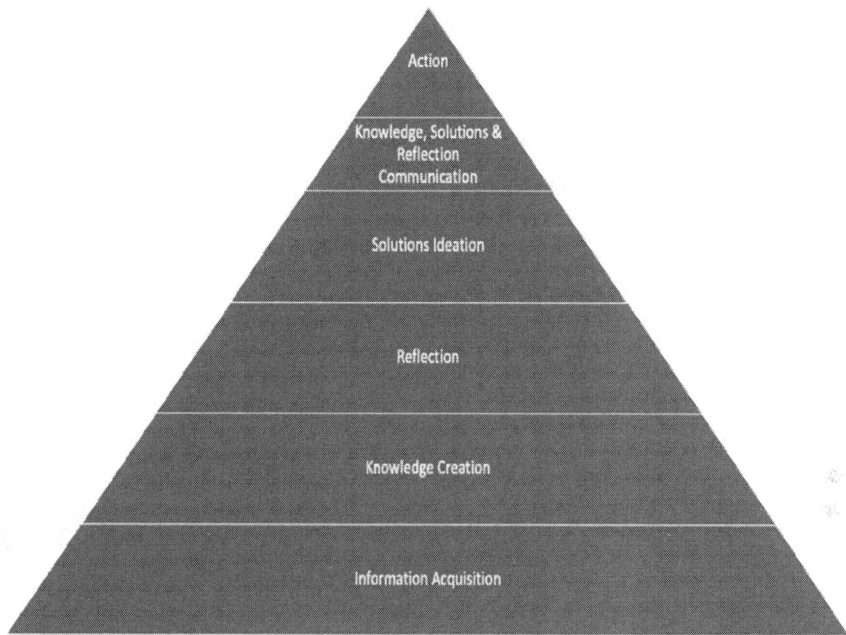

FIGURE 1.3   Hierarchy of Student-Created Media Potential

Reprinted with permission from American Library Association. Image from Scott Spicer, *Student-Created Media Designing Research, Learning, and Skill-Building Experiences* (Chicago: ALA Editions, 2022).

This requires research and enables the students to be well-informed about what they discuss. It also makes a topic as expansive as climate change writ large, less overwhelming and more manageable for both the media project coverage and topic understanding. Three, it is important that students consider current and ideate potential future solutions. This enables them to determine how best to communicate that message multimodally, using a full palate to communicate their ideas. In addition, solutions identification and ideation move the issue from course provocation to situating the students as potential change agents, whereby students think to themselves, "I may actually be able to do something about this problem."

We now want to illustrate just how these principles work in practice.

Case Study #1—Ecosystem Health Grand Challenges Curriculum: Video Investigation and Solutions Modeling Media Project Related to Ecosystem Health

There are multiple ways to design a curriculum that incorporates both the personal reflective and multi-perspectival components. For example, the Ecosystem Health Grand Challenges Curriculum course at the University of

Minnesota-Twin Cities includes a student-created media assignment designed to situate the students as change agents on an issue related to Ecosystem Health.

In order to accomplish this goal, the student media assignment requires students to:

a Provide the background on a topic related to sustainability.
b Identify key stakeholders on the issue (multi-perspectival) with a systems map on how these stakeholders are interconnected.
c Propose a blue-sky solution for the topic.
d Consider and describe the unintended consequences of their proposed solution.

Incorporating all these required components into a single media project is a heavy lift both on technical production and on curriculum scaffolding. Students need to learn about how to conduct a stakeholder analysis while they are producing their systems map infographic, later incorporated into their video project—the nutrients of this project ensure an immersive learning experience on the topic that the students will not forget.

Created Media: Designing Research, Learning, and Skill-Building Experiences demonstrate this potential (Spicer, 2022c, pp. 99–109). First, students describe the impact of the fast fashion industry's harmful environmental and human rights abuses. They set up the issue by providing researched background and using interviews with students regarding how much they would pay as a primary concern. They continue to describe the various stakeholders involved in the fast fashion supply chain and consumer ecosystem, using a systems map they created for illustration. They propose blue-sky solutions, including an industry ethics rating system, a tax offset program dependent on rating, and social media influencer campaign to promote compliance. Finally, they recognize governmental regulation's role in holding these industries accountable. The students leverage an array of mixed media, including filmed interviews, third-party media, and voice to communicate their case study. Evidence of subject knowledge understanding, critical thinking, and creative problem solving is evident throughout the video, their final class in-person presentations, and an interview with one of the students. Finally, some transformative effects are present; for example, one student referenced their project in a sustainability-related grant, another referenced their project on their graduate school application and two others went on to work at locations related to those they profiled in their video projects. This outcome is at the apex of the Hierarchy of Student-Created Media Potential. Concerning audience uptake, Spicer has re-used this video in other courses to demonstrate a quality student-created media project; it is on the *Student Produced Sustainability Project Gallery z.umn.edu/sustainabilityinvestigationsolutionsvideo* and also on the Grand Challenges Curriculum course website designed for public communication *gcc.umn.edu/node/701.*

In terms of composition, the reuse of media with intentionality is important. From a media and visual literacy perspective, the instructors and media support professionals are careful to remind the students that the public is one of the potential audiences for their work, pending student approval to share openly. Some of the folks viewing their media projects may disagree with their recommended solutions. In this event, not only does the research evidence supporting their position need to be accurate—see tier one on Spicer's Hierarchy of Student-Created Media—but the images also need to match the narrative. For example, having a stock image of one locale that does not match up with the case study being described is not good enough. As Buturian suggests, "Educators who require images as part of academic work soon find the contradiction that students who have been weaned on images, and who have been described as visual learners, are not necessarily educated in visual literacy" (Buturian, 2016, Chapter 3, Scaffolding Exercises section, par. 8).

Rhetorically, if the image or sound use is incongruent, detractors will use their misuse of media to discredit their entire argument. As we point out in our instruction, this feedback is not only related to the project, but is also an occupational and civic critical media literacy skill set that we are trying to develop as students make their way into the world after graduation.

By shaping the pedagogy through a critical media lens, the Ecosystem Health Grand Challenges Course instructors and related support professionals are blending the media arts and critical media literacy traditions. This immersive experience leads to transformative effects. For example, one student began to sew her own clothes due to learning about the harmful human rights and environmental effects of fast fashion (Spicer, n.d.). Additional information on the specifics of this project with actual student examples can be found in the *Student Produced Sustainability Project Gallery z.umn.edu/sustainabilityinvestigationsolutionsvideo*.

Case Study #2—Water, Water, Water Everywhere: Personal Narrative Digital Stories Related to Water Use

Another effective way to close the loop is to use the media project for personal reflection that situates the students in relationship to the topic at hand. As noted earlier, developing stories is a compelling way to immerse students in learning. That said, if one is to impact change, it is not only important to have a deep appreciation of one's own perspective. It is just as vital to recognize alternative points of view that may not be surfaced in this style of media creation. Fortunately, there are multiple curricular strategies that can be deployed to "close the loop."

For example, in the case of the Water, Water, Everywhere course taught by Linda Buturian at the University of Minnesota (see Chapter 6 in this book for a description of those courses), students are encouraged to reflect and communicate via media about their relationship with water. In order to be exposed to multiple perspectives, students viewed other student works and sometimes also had subject experts profiled in their videos in attendance at the final screening event. These conversations—when paired with a screening or listening (podcast) session—are rich and add an amount of depth and intimacy to the topic that lasts not only with the student that interviewed this individual in their media project but also with the other students in the class (in addition to instructors and related support professionals and visitors).

With respect to media composition for a personal narrative digital story, these videos often include subject-matter materials, combined with personal images and video clip interviews. Just as the Grand Challenges Course required students to create a systems map infographic of topic stakeholders prior to creating their video, early on in the semester, Buturian required her students to create an infographic representing their thoughts on water sustainability in a visual form. These lighter-weight exercises help students develop comfort with media making and provide learning space for rhetorical communication value of these productions when shared with the class. Additional information on the project specifics with actual student examples can be found at Student Produced Sustainability Project Gallery *z.umn.edu/sustainabilitypersonalnarrativevideo*

Case Study #3—Sustainability Public Policy Video Investigation Media Project

While the Ecosystem Health Course assignment described incorporates multiple components, a comparatively less creation-intensive video investigation media assignment can also be impactful. In Energy, Sustainability and Water policy classes taught through the University of Minnesota's Humphrey School of Public Affairs, students in groups create videos that explore sustainability-related topics in-depth, communicating a range of perspectives. These videos are stylistically informational to educate the audience (e.g., journalists, the general public, or policymakers) on an issue with potential options.

Notably, in some instances, these videos do not necessarily advocate for a particular position. This neutrality opens up a space to keep an open mind and debate the merits of the works presented (a core value in public policy discourse). In some classes, students also write a policy brief document that provides the individual student's point of view with evidence to support their argument. Finally, students typically participate in a final end-of-semester screening, where their responses to audience member inquiries further illustrate their knowledge of their learning during the Q&A portion of their presentation.

Regarding composition, these videos generally include mixed media found on the web or campus-licensed stock footage, combined with recorded subject expert interviews and voice-over narration. Given the scientific nature of these presentations, significant technical knowledge of the subject matter is essential to develop the appropriate interview questions and describe their case studies.

Concerning transformative effects, these videos demonstrate to future employers the subject knowledge students have obtained while also providing evidence of their ability to communicate complex issues in an accessible manner to various audiences. Additionally, in some cases, these videos have been shared publicly on social media, including YouTube and Facebook, and contests. The Student Produced Sustainability Project Gallery showcases some of these projects with commentary: *z.umn.edu/sustainabilityvideoinvestigation*

Conclusion: Youth-Created Media and Activism

A central question addressed in this book is whether youth-created media projects on climate change topics necessarily result in activism. Of course not. Media projects, like all assignments, are at least initially limited to the confines of the classroom context. That said, when designed with intentionality—and, even better, in partnership with community partners—these projects allow students to walk away with a deeper understanding of at least one issue related to climate change. And with the knowledge to effect change. The reflection, exposure to multiple perspectives, project sharing, and potential relationships gained through this process equips the students with the skill sets and motivation resulting in transformative effects that will endure long after the projects are complete. Beyond their own contributions, through this learning, students will be encouraged to support policies and politicians that support climate action and adaptation initiatives so they can take back their future.

References

Anderson, B. (1983). *Imagined communities: Reflections on the origin and spread of nationalism.* Verso.

Beach, R., Share, J., & Webb, A. (2017). *Teaching climate change to adolescents: Reading, writing, and making a difference.* Routledge.

Bell, A. (2021). *Our biggest experiment: An epic history of the climate crisis.* Counterpoint.

Boyd, B. (2009). *On the origin of stories: Evolution, cognition, and fiction.* Belknap Press.

Burg, N. (2022, April 20). Who *funds* the fight *against* climate change? *Means & Matters.* https://meansandmatters.bankofthewest.com/article/sustainable-living/taking-action/who-funds-the-fight-against-climate-change/

Burney, M. (2020, July 2). In New Jersey schools, climate change education will be mandatory. *The Philadelphia Inquirer.* https://www.inquirer.com/education/climate-change-new-jersey-mandatory-curriculum-tammy-murphy-20200702.html

Buturian, L. (2016). *The changing story: Digital stories that participate in transforming teaching & learning.* College of Education and Human Development, University of Minnesota. https://www.cehd.umn.edu/the-changing-story/

Clayton, P., Archie, K. M., Sachs, J., & Steiner, E. (Eds.) (2021). *The new possible: Visions of our world beyond crisis*. Cascade Books.

Crist, E. (2013). On the poverty of our nomenclature. *Environmental Humanities, 3*, 129–147.

Crist, E. (2019). *Abundant earth: Toward an ecological civilization*. The University of Chicago Press.

Emmett, R. S., & Nye, D. E. (Eds.) (2017). *The environmental humanities: A critical introduction*. The MIT Press.

Ghosh, A. (2016). *The great derangement: Climate change and the unthinkable*. University of Chicago Press.

Ghosh, A. (2021). *The nutmeg's curse: Parables for a planet in crisis*. University of Chicago Press.

Haiven, M. (2014). *Crises of imagination, crises of power: Capitalism, creativity, and the commons*. Fernwood Publishing.

Harari, Y. N. (2015). *Sapiens: A brief history of humankind*. HarperCollins.

Herman, D. (2002). *Story logic: Problems and possibilities of narrative*. University of Nebraska Press.

Herman, D. (2013). *Storytelling and the sciences of mind*. MIT Press.

History of Education for Sustainable Development. (n.d.). Retrieved from https://unescochair.info.yorku.ca/history-of-esd/

History of Environmental Education. (n.d.). Retrieved from https://www.k12academics.com/education-subjects/environmental-education/history

Hogan, P. C. (2003). *Cognitive science, literature, and the arts: A guide for humanists*. Routledge.

Hogan, P. C. (2009). *Understanding nationalism: On narrative, cognitive science, and identity*. The Ohio State University.

Hogan, P. C. (2011). *What literature teaches us about emotion*. Cambridge University Press.

Hogan, P. C. (2013). *How author's minds make up stories*. Cambridge University Press.

IMF. (2022). Fossil Fuel Subsidies. https://www.imf.org/en/Topics/climate-change/energy-subsidies

International Panel on Climate Change. (2018). *Summary for policymakers*. Cambridge University Press. doi:10.1017/9781009157940.001

Johnson, A. E., & Wilkinson, K. K. (Eds.) (2020). *All we can save: Truth, courage, and solutions for the climate crisis*. One World.

Joseph, P. (2017). *The new human rights movement: Reinventing the economy to end oppression*. BenBella Books.

Kimmerer, R. W. (2013). *Braiding sweetgrass: Indigenous wisdom, scientific knowledge, and the teachings of plants*. Milkweed Editions.

King, J. B., Todd Whitman, K., Duncan, A., Jewell, S., & McCarthy, G. (2020, November 18). Open letter to president-elect Joseph Biden vice president-elect Kamala Harris. https://www.k12climateaction.org/img/Biden-Transition-ED-Climate-Formatted.pdf

Klein, N. (2014). *This changes everything: Capitalism vs. the climate*. Simon and Schuster.

Klein, N. (2019). *The (burning) case for a green new deal*. Simon and Schuster.

Klein, N. (2021). *How to change everything: The young human's guide to protecting the planet and each other*. Anatheum Books.

Lent, J. (2017). *The patterning instinct: A cultural history of humanity's search for meaning*. Prometheus Books.

Lent, J. (2021). *The web of meaning: Integrating science and traditional wisdom to find our place in the universe*. New Society Publishers.

López, A. (2021). *Ecomedia literacy: Integrating ecology into media education*. Routledge.

Mann, M. E. (2021). *The new climate war: The fight to take back our planet*. Public Affairs.

Marshall, G. (2014). *Don't even think about it: Why our brains are wired to ignore climate change*. Bloomsbury.

Monbiot, G. (2016). *How did we get into this mess? Politics, equality, nature*. Verso.

Monbiot, G. (2017). *Out of the wreckage: A new politics for the age of crisis*. Verso.

Moore, J. W. (Ed.) (2016). *Anthropocene or capitalocene: Nature, history, and the crisis of capitalism*. PM Press.

Morton, T. (2013). *Hyperobjects: Philosophy and ecology after the end of the world*. The University of Minnesota Press.

NGSS, Next Generation Science Standards. (2013). Appendix E–progressions within the next generation science standards. https://www.nextgenscience.org/sites/default/files/resource/files/AppendixE-ProgressionswithinNGSS-061617.pdf

Nixon, R. (2011). *Slow violence and the environmentalism and of the poor*. Harvard University Press.

Norgaard, K. M. (2011). *Living in denial: Climate change, emotions, and everyday life*. MIT Press.

Oreskes, N., & Conway, E. (2010). *Merchants of doubt: how a handful of scientists obscured the truth on issues from tobacco smoke to climate change*. Bloomsbury.

Orr, D. W. (2004). *Earth in mind: On education, environment, and the human prospect*. First Island Press.

Overland, I., & Sovacool, B. K. (2020, April). The misallocation of climate research funding. *Energy Research and Social Science, 62*. doi:10.1016/j.erss.2019.101349

Oziewicz, M., Attebery, B., & Dĕdinová, T. (Eds.) (2022). *Fantasy and myth in the anthropocene: Imagining futures and dreaming hope in literature and media*. Bloomsbury Academic.

Oziewicz, M. (2015). *Justice in young adult speculative fiction: A cognitive reading*. Routledge.

Oziewicz, M. (2022). Fantasy for the anthropocene: On the ecocidal unconscious, planetarianism, and imagination of biocentric futures. In M. Oziewicz, B. Attebery, & T. Dĕdinová (Eds.), *Fantasy and myth in the anthropocene: Imagining futures and dreaming hope in literature and media* (pp. 58–69). Bloomsbury Academic.

Oziewicz, M. (2022a). Planetarianism now: On anticipatory imagination, young people's literature, and hope for the planet. In M. Paulsen, J. Jagodzinski, & S. M. Hawke (Eds.), *Pedagogy in the anthropocene: Re-wilding education for a new earth* (pp. 241–256). Palgrave.

Oziewicz, M. (2022b, January 14). Why children's stories are a powerful tool to fight climate change. *YES! Magazine*. https://www.yesmagazine.org/opinion/2022/01/14/climate-change-childrens-stories

Oziewicz, M., & Saguisag, L. (2021). Introduction: children's literature and climate change. *The Lion and the Unicorn, 45*(2), v–xiv.

Panos, A., & Damico, J. (2021). Less than one percent is not enough: How leading literacy organizations engaged with climate change from 2008 to 2019. *Journal of Language and Literacy Education, 17*(1), 2–21.

Paulsen, M., Jagodzinski, J., & Hawke, S. M. (Eds.) (2022). *Pedagogy in the anthropocene: Re-wilding education for a new earth*. Palgrave.

Read, R., & Alexander, S. (2019). *This civilization is finished: Conversations on the end of the empire—and what lies beyond.* Simplicity Institute.

Schank, R. C. (1995). *Tell me a story: Narrative and intelligence.* Northwestern University Press.

Shiva, V., & Shiva, K. (2021). *Oneness vs the 1%: Shattering illusions, seeding freedom.* Chelsea Green Publishing.

Siperstein, S., Hall, S., & LeMenager, S. (Eds.) (2017). *Teaching climate change in the humanities.* Routledge.

Spicer, S. (2022a). *Student-created media designing research, learning, and skill-building experiences* (pp. 5–7). ALA Editions.

Spicer, S. (2022b). *Student-created media designing research, learning, and skill-building experiences* (pp. 8–11). ALA Editions.

Spicer, S. (2022c). *Student-created media designing research, learning, and skill-building experiences* (pp. 99–109). ALA Editions.

Spicer, S. (n.d.). *Student Produced Sustainability Project Gallery.* z.umn.edu/sustainabilitymediaprojects

Supran, G., & Oreskes, N. (2021). Rhetoric and frame analysis of ExxonMobil's climate change communications. *One Earth, 4*, 696–719. doi:10.1016/j.oneear.2021.04.014

Szeman, I., & Boyer, D. (2017). Introduction: On energy humanities. In I. Szeman & D. Boyer (Eds.), *Energy humanities: An anthology* (pp. 1–14). John Hopkins University Press.

Taylor, C. (2013). The discourses of climate change. In T. Cadman (Ed.), *Climate change and global policy regimes* (pp. 17–31). Palgrave Macmillan.

The essential principles of climate literacy. (2009). National Oceanic and Atmospheric Administration. https://www.climate.gov/teaching/climate

The Red Nation (2021). *The red deal: Indigenous action to save our earth.* Common Notions.

Turner, M. (1996). *The literary mind.* Oxford University Press.

UN Intergovernmental Panel on Climate Change. (2018, October 8). Special report, global warming of 1.5° C. https://www.ipcc.ch/sr15/

Wheal, J. (2021). *Recapture the rapture: Rethinking god, sex, and death in a world that's lost its mind.* HarperCollins.

Young, R. (Ed.) (2022). *Literature as a lens for climate change: Using narratives to prepare the next generation.* Rowman & Littlefield.

Zimmerman, L. (2022, February 7). Our climate discourse is gradually normalizing an atrocity. *Current Affairs.* https://www.currentaffairs.org/2022/02/our-climate-discourse-is-gradually-normalizing-an-atrocity

2

CENTERING UTOPIA

Fostering Youth Climate Change Education by Exploring and Envisioning Hopeful Futures

Michelle E. Jordan, Catherine Lockmiller, and Steven J. Zuiker

Introduction to Imagining the Future

Utopias, according to dictionary definitions, imply near-perfect alternatives to the present; yet, near perfection often remains implausible and unrealistic. In contrast, the science fiction author Kim Stanley Robinson (2020) views the idea of utopia not as "a perfect-end-state society" but "as a progressive movement in history, with each generation doing better than the generation before, in substantial ways, in terms of equality, justice, and sustainability." We join authors like Robinson in arguing that utopias may be not only useful, but urgently necessary.

> ...the onrush of catastrophic climate change has forced a reckoning. We either invent and institute a better way, or a mass extinction will take us down with it. Necessity has thus jammed utopia into history and turned it from a minor literary genre into an important tool of human thought. We need it like never before.
>
> *(Robinson, 2016)*

There is utopian thinking or the apocalypse when staring down a future loaded with climate catastrophe. We opt to stay with utopia in fostering climate change literacy with youth. We contend that providing opportunities to explore the utopias articulated by others and envisioning their own imagined future utopias can help youth navigate climate change, enabling hopeful action. Concerning climate change education, we explore how to combine the processes of exploring and envisioning utopias by coupling information literacy (for exploration) with futures thinking (for envisioning).

DOI: 10.4324/9781003335276-4

Drawing from the diverse scholarship related to information literacy, futures thinking, and utopia as a literary genre, we argue that utopian thinking is key to climate change education. Engaging with utopian genres frames pathways for students to critically explore an ecologically secure future with peers. Further, opportunities for youth to explore their relationship to expansive literary genres of utopian, dystopian, and apocalyptic media provide possibilities for leveraging creative media to exercise agency in relation to the civic-present and the civic-futures that youth envision for themselves and their communities. However, utopian thinking and its attendant opportunities depart from conventional approaches to sustainability education.

Some teachers, club leaders, parents, and other facilitators inevitably foster youths' information literacy related to climate change using only canonical information sources and transmission-based teaching about the science of climate change. But dissemination alone is not enough. Information-focused pedagogies emphasizing conceptual knowledge are insufficient to prompt knowledge-based climate change action (Trott, 2020). This suggests that action is desperately needed as citizens and nations remain short of adequate response to impending climate crises (Turrentine, 2022). Mitigating and adapting to climate change will require not only information but also emotional labor (Moser, 2007), imaginative capacity (Hicks, 2014), and collective/civic dialogue (Ojala, 2017). Moreover, this chapter investigates youth's collective ability and will to imagine novel tomorrows for building solidarity to protect and project shared ecologies.

This chapter explores the conjecture that centering climate change education on utopian thinking coupled with information literacy and futures thinking can help educators support youth in exploring and envisioning hopeful futures for themselves and their communities. Moreover, doing so can help develop a constructively critical lens toward climate change issues (Beach et al., 2020) and foster constructive hope necessary to cultivate and sustain civic imagination and social action (Siegner & Stapert, 2020). Here, *information literacy* is broadly defined to include its close relatives, media literacy, and digital literacy, and includes engagement with a wide range of utopian creative media forms (e.g., books, videogames, films, music, and dance). Engaging in these media enables youth to read the world and words and to write to produce them (Kafai & Peppler, 2011). In addition, these forms can accompany and arguably enhance engagement with scholarly and academic information sources. Although information science, and therefore information literacy, has focused on written documents, anything can be an informative *media object* insomuch as its purpose is to intentionally communicate meaningful messages (Crane, 2001).

This chapter unfolds in three parts to develop our investigation of the conjecture above. First, we review literature informing our approach to centering utopia in climate change education by coupling information literacy and futures thinking. Second, we present an approach inspired by this literature and describe enacting a series of information literacy workshops for high school students that embody the approach. Specifically, we concentrate on how workshop activities outlined pathways for exploring scholarly and academic information sources and utopian creative media to support youths' climate change education. Finally, this chapter proposes ways K–12 educators can design learning opportunities that invite youth into the collective work of imagining and developing hopeful futures.

A History of Utopia

The concept of "utopia" can be traced back to 16th-century lawyer and rhetor, Thomas More. In *Utopia*, More's narrator describes an Edenic island nation called *utopia*. The name itself is an Anglicized portmanteau of the Greek words *topos* (meaning "place") affixed to a suffix that combines *ou* ("no") and *eu* ("good"). Thus, "utopia" is paradoxically both a "good place" and "no place." As a description of a supposedly perfect society, More's Utopia resembles other mythical and philosophical places. At the same time, *utopia's* impact on political literature has been so immense as to establish entirely new literary genres. Importantly, *utopia* has not only culminated in a literary form; it has also given rise to centuries of political philosophy that seeks to answer key questions concerning utopia: what does utopia look like, and can it be made real?

Ruth Levitas (1990) outlines this historical discourse as a "socially constructed response to an *equally* socially constructed gap between the needs and wants generated by a particular society and the satisfactions available to and distributed by it" (p. 210). As a result, Levitas conceptualizes utopia as a (potentially) playful and creative space wherein a societal lack can be approached, diagnosed, and treated with the desires which would serve to assuage what is lacking. Utopia is therefore a *method* which Levitas (2013) describes as a "conjuring of its form…of a process moving between openness and closure" (p. 220). Further, as a literary form, utopia functions as a teaching tool that also conveys an imaginary wherein the participant can feel their way through the space, considering its architecture, ethics, and mechanical operations (Levitas, 2013). From this framing, we can gather that there is a didactic—or educational— quality to utopia as a mode. It provides access for learners to partake in critical reflection and further entails the potential for co-creation (Giroux, 2007). More's *utopia* consists of this quality. It provokes the reader to question the viability of its supposedly perfect political states. Further, it establishes theoretical goalposts that can be used as measuring rods against the reader's present political condition.

The Utopian Present

Recent literary texts invite the same responsivity as *utopia and other early utopian texts*. One example is Kim Stanley Robinson's 2020 science fiction novel, *The Ministry for the Future*. This novel is explicitly concerned with global warming and environmental collapse. It takes seriously the problems inherent in present-day neoliberal, market-driven social structures which prevent the enactment of meaningful climate action. From a didactic perspective, it presents a utopian imaginary wherein readers can come to terms with and ultimately unravel the systems tied directly (and indirectly) to climate catastrophe.

At the same time, Margaret Atwood's (2003) science fiction novel, *Oryx and Crake*, presents a *dystopian* space wherein runaway scientific experimentation and a fully deregulated free market intermingle and facilitate the mass extinction of the human species. Even though *Oryx and Crake* presents a dystopia, the intent is arguably still utopian. Dystopias convey a warning or series of warnings that necessitate massive change, the violent upheaval of social structures causing present harm, and unconstrained ones that will lead to horrific consequences (Sargent, 1994). Levitas (2013), among others, has argued that dystopia facilitates a dead-end view of the future, critiquing dystopia as a contrivance of a neoliberal marketplace in which negative depictions of violent social structures can be commodified. However, Levitas indicates that dystopia can posit a similar playful imaginary as utopia. In this manner, it is *not* anti-utopian. Anti-utopianism is a specific modality that refuses utopian imaginaries (Sargent, 1994). Instead, utopian-esque dystopias, such as *Oryx and Crake*, invoke "another possible future" and maintain a method by which that future can be made to happen (Levitas, 2013).

Perhaps the best way to understand utopia and dystopia in this regard is to consider their capacity to call people to action. In other words, utopias that invite participation and critical exploration can produce positive change. Engagement with utopian media objects is aligned with shifts away from climate change education as simply dispensing information toward action-oriented pedagogies (Walsh & Cordero, 2019). Regardless of the way it is packaged, utopia (or dystopia) can be understood as (a) a framing for describing a better political place that is not yet present, (b) a method for exploring a societal lack and considering how best to treat it, (c) a didactic framing that invites participants into an imaginary space, and (d) a call to co-creative action. These four criteria align with our vision of a pedagogical space that invites learners to imagine large-scale structural change leading to a better world, one that is not yet realized but cannot be written off as untenable. Further, these criteria establish utopia as a meaningful learning device. Through utopian framing, educators can spark difficult conversations about the world as it is now and how a better world—a renewed world—can be informed, imagined, and created. In other words, pedagogical investments in utopian orientations rely not on constraining discussion to only optimistic outcomes; instead, they aim to inspire critical exploration and "thoughtful optimism" that is closely aligned with hoping for and working toward a better future (Finn & Wylie, 2021). Thus, even engagement in dystopic forms of utopian genres can offer opportunities for imagining hopeful futures.

We see the role of hope as paramount to ecosystem and social renewal and utopia as a transformative teaching tool for nurturing hope. Research suggests that globally youth largely hold fatalistic and apocalyptic orientations toward climate futures (Angheloiu et al., 2020; Wright et al., 2021). Pervasive pessimism can foster a sense of hopelessness, inhibiting agency and thwarting individual and collective action (Ojala, 2017). For these reasons, a steady diet of media that turn toward the dystopian, post-apocalyptic, and anti-utopian may undermine youths' ability to imagine counter-narratives to climate catastrophe. Dystopia is not enough. Advocating the need for utopia in dark times, Giroux (2007) identified a turn toward utopia as important for catalyzing agency and "educated hope" (p. 21). Educated hope avoids both the pitfalls of "artificial optimism" (p. 34) and fatalistic pessimism that orients to climate change collapse as inevitable and inescapable. Both poles function to depoliticize, disenfranchise, and disempower people in regard to entangled social, economic, and environmental concerns rather than offering critical engagement with civic and everyday life offered by a *pedagogy of hope* (Freire, 2021).

Utopia and Climate Change

There is a rich tradition of applying utopian methods to environmental activism (McKnight, 2020). Even in *Utopia,* More analyzed human relationships with the rest of the natural world. However, as a Christian humanist, this occurs through a religious and sacral lens (Sabry, 2021) rather than through a scientific and human progressive lens as expressed by Robinson (above). More famously, the British socialist and leader of the early Arts and Crafts Movement, William Morris, funneled the anxieties of industrial England, worker exploitation, and expropriation from nonhuman ecological spaces into his literary work, namely *News from Nowhere* (Davidson, 2019). Morris's work, in particular, facilitated— or at least predicated—a 20th-century literary subgenre combining ecological utopian goals with genre literature (i.e., science fiction and fantasy).

Post-apocalyptic fiction is itself a distinct subgenre in its own right. Utopian ecological literature, more broadly, is indicated in climate-focused texts across media forms. These forms include titles as varied as *Black Panther* (Coogler, 2018, feature film), *Nausicaa of the Valley of the Wind* (Miyazaki, 2012, graphic novel), *Parable of the Sower* (Butler, 1993, print book), and *Star Trek* (Roddenberry, 1966–1969, televised series), among others. For example, the video game

Endling: Extinction Is Forever (Herobeat Studios, Developer, 2022) is an ecological, dystopian narrative that focuses on the life of a mother fox protecting her young from humans who are starving due to mass extinction caused by climate change.

The media mentioned above challenge the necessity of centering didacticism within the utopian frame. Instead, they provide learners with the imaginary worlds contained within their narratives. Rather than adopting a didactic orientation to climate change education, readers are invited to develop a richer understanding of earth's ecological present. They can also conceptualize the directionality necessary to move forward into an ecologically sustainable future.

Germane to climate change education is technological innovation's role in exasperating and mitigating climate catastrophe. Utopian literature has often been concerned with technologies and how they produce utopia or prevent it. This literature includes late 19th- and 20th-century utopian texts such as *Looking Backward* (Bellamy, 1888) and *Herland* (Gilman, 1915/1998). Each of these texts describes utopian states that rely on some level of technology in order to function. For example, *The Time Machine* (Wells, 1895) presents a dystopian future that conveys increased exploitation and the eventual ruination of proletarian classes through technological modes.

Notably, technologies also function as sources of critique within utopian texts. *Oryx and Crake* (mentioned above) is a prime example of this technique since it is focused on biotechnologies and genetic experimentation that cause runaway environmental destruction. As a result, utopian literature is particularly well situated to wrestle with climate change and technologies that contribute to both harmful climate change and technologies that can help resolve or mitigate human-caused climate change. Further, utopian literature pinpoints the irreducible connection between technologies and the societies in which they function. Therefore, by crafting learning initiatives around climate-focused utopian media, teachers can invite learners to consider technologies' economic, environmental, and social implications within imaginary worlds. By extension, students can reflect on those technologies' roles in their neighborhoods and communities.

Centering Utopia in Information Literacy for Climate Change Education

Information literacy has long been intertwined with managerial and neoliberal conceptualizations of knowledge production (Seale, 2016). This is particularly visible within librarianship, where information literacy is a key pedagogical project, and where neoliberal praxis dictates learner interactions with concepts of information exchange (Coleman & Pankl, 2020), often relying on deficit-based and context-flattening models. Most famously, the Association of

College & Research Libraries' *Framework for Information Literacy for Higher Education* is a focal document that conveys the struggle to resist neoliberal pedagogy while retaining the structural currents that facilitate neoliberalism in the first place (Seale, 2016). The framework provides a means for teachers to convey the power relations that result in knowledge becoming commodified, but fails to move beyond recognition and basic critique. Librarians and other teachers who rely on it are beholden to ways of thinking about information literacy which remain neoliberal and which cannot be construed as utopian.

Similarly, the academic publishing ecosystem from grades K to 20 is beholden to power structures that restrict revolutionary and utopian thinking, and instead center incremental, "evidence-based" pedagogical approaches. Such approaches simultaneously constrain creative choices and anti-institutional aesthetics, including utopian media. This results in a double bind for educators who wish to produce utopian social imaginaries in a climate change education environment. Teachers cannot use exemplary, evaluative toolkits, nor are they likely to find the information resources necessary to provoke utopian thinking in academic discourses. Therefore, we reject the sole use of traditional information literacy methods, foregrounding creative media to establish utopian dialogue in climate change education environments.

We envision a learning environment wherein utopia works hand-in-hand with information literacy to provoke learners into considering hyper-alternative solutions to climate change. This entails a multi-step process that begins with learning how to retrieve standard, evidence-based information, identify utopian media that deals with similar concepts to compare how they align and differ, and finally, imagine utopian futures that coalesce current technologies, emerging technologies, civic structures/practices, and utopian ideals.

Centering Utopia in Futures Thinking for Climate Change Education

Diverse youth networks spanning the globe mobilize around climate action (Trott, 2021) and activism (Sunrise Movement, 2023). Youth are in a unique position to take transformative roles in envisioning, enacting, and embodying utopian futures, perhaps because they have more to lose in a future where humans have failed to mitigate climate change and less at stake in the present when resisting the status quo (Salemink et al., 2020). By participating in such activism, youth cultivate their optimistic orientations to the future (Cattell, 2021) and exert a measure of control over the trajectory of their own lives (Salemink et al., 2020). Their efforts also result in collective outcomes like shared interests and solidarity among youth (Jenkins et al., 2020). Moreover, working together to imagine how real places can change, youth transcend current social situations, often evoking utopian elements unrealized in the past or present (Salemink et al., 2020).

The climate action and activism above are promising. However, more support for youth is necessary to accelerate and amplify efforts. These efforts include raising awareness and prompting collective action, but also affecting political discourse using sustainable and socially just eco-imaginaries (Epstein-HaLevi et al., 2021). These supports can also counterbalance current educational systems that may inadvertently exacerbate apathy and inertia among youth. Specifically, when climate change education treats the future as predetermined, it positions youth as passive recipients of imminent change rather than active producers whose actions resist predetermined futures and imagine (Holfelder, 2019) radically hopeful futures instead.

In order to provide support and counter a single future, we center utopias when engaging youth in futures thinking. Centering utopias is a pedagogical means to couple futures thinking with information literacy using utopian genres as a common frame and, in turn, as generative space for youth to envision and shape hopeful climate change futures across neighborhood, community, and global scales. In recent decades, futures thinking has emerged in response to the perceived need for more democratized and pluralized participation in shaping emergent civic and social futures (Gidley, 2017). Futures thinking is a method for inviting participation and critical exploration that is in keeping with youths' desires to contribute to climate action. It also complements utopian methods. Much as Levitas (2013) described utopia as a method for moving between openness and closure, futures thinking processes invite participants to adopt a view of the future as open and shapeable (Jordan et al., 2021), thereby fostering agency for coupling imagination and action to cultivate visions of hoped-for futures. Like utopian methods, futures thinking seeks to enhance imaginative capacity, putting imagination to work to envision ways to create conditions, define processes, and shape outcomes related to preferred futures (Finn & Wylie, 2021).

Finally, centering utopia in futures thinking builds upon a common foundation in narrative and storytelling. Research has shown that storytelling can effectively promote climate change education among youth, fostering comprehension, agency, and voice (Littrell et al., 2020; Trott, 2020; Walsh & Cordero, 2019). Thus, futures thinking processes can center youth's engagement with utopian storytelling across various media forms, as well as the production of their own utopian creative media through which they imagine preferred future social and technological futures for themselves, their communities, and the planet.

Coupling Utopian-Centered Information Literacy and Futures Thinking: An Illustrative Example

This section develops an illustrative example of climate change education that couples utopian-centered information literacy and futures thinking. Our example is drawn from an ongoing program of design-based research focused on the climate change education topic of post-carbon energy transitions. The 21st century will see the dissolution of carbon-based energy systems and a transition to renewable energy systems (Pasqualetti, 2021). Solar innovations will be central to renewable energy systems but they will also remain intertwined with social, political, economic, ecological, and ethical considerations (Miller, 2022). Exploring and envisioning solar futures therefore entail not only what new solar technologies might emerge, but also how communities might develop, distribute, own, and utilize them (Jasanoff & Kim, 2015).

Against the general backdrop of energy transitions, our example features three workshops designed for and enacted with youth. Echoing the conjecture above, our intention in designing the workshops was to support youth in exploring and envisioning hopeful energy futures. Specifically, we wanted to understand how coupling utopian-centered information literacy and futures thinking positioned youth to exercise agency through the co-development of solar innovations (investing in the present) and solar narratives (exploring the possible) as contributions to ecologically sustainable transitions to post-carbon energy systems.

All workshops were embedded in a six-week summer solar energy research experience program (220 hours in total duration). This program brought together a small group of US high-school-aged youth (13–17 years old) with strong ties to the same Latino/a community. In general, the summer program situated youth as agents of change. For example, youth were imbued with the ability to simultaneously learn and contribute to impending renewable energy system transitions in their shared community by collaborating with a diverse network of adult facilitators. Their shared goals were two-fold: (a) develop community-based solar energy innovations and (b) inspire others by producing utopian solar futures narratives. The featured workshops (~ten hours) were one of five complementary learning strands featured in the program.

The design and instantiation of the workshops emphasized youth's engagement with information literacy through utopian creative media, with a secondary emphasis on scholarly research. We see both types of information sources as rich media objects with unique knowledge possibilities to inform climate change education. At the same time, each offers distinct possibilities for thinking about the future. Specifically, scholarly research provides insight into what is already present within the social imaginary or is presumed to be imminent in the near future. Thus, using traditional information search and evaluation strategies to find answers to particular questions can immediately inform practical and critical uses of near-term innovations. In contrast, engaging with utopian creative media establishes a space to narrativize and imagine new possibilities and social landscapes to which intellectual, political, and creative energies can be directed.

Utopian creative media does not have a precise goal of finding and spreading information and yet it can still contribute to that activity. Its purpose is to be in conversation with the reader or the viewer—not to tell something to the reader or the viewer. Thus, regarding our workshops, we attempted to give the learners space to engage with utopian creative media in order to expand beyond typical research paper report activities. We wanted the youth to interact with a wide range of information sources in multiple media forms and bring their own experiences to the fore in that interaction. Finally, by coupling utopian genres and scholarly resources, we hoped to bring together reason and imagination to prompt speculation (Pendleton-Jullian & Brown, 2018) needed to support people in constructing alternative futures that are both plausible and desirable (Bai et al., 2016). More specifically, we wanted to bring hopeful elements of existing and near-future socio-technical innovations into conversation with radically transformative far-future civic and social processes for promoting sustainable human-environmental relationships.

The activities that comprised the featured workshops were co-designed by Michelle, Catherine, and Steve, and hosted by Catherine, a librarian with a background in utopian literature. Together, workshops focused on the concept of *renewal*, a nod to the larger program's overall focus on renewable energy transitions and also the idea of utopian thinking as a catalyst for social and political transformation. Each workshop was spread across multiple sessions and modes. The workshop host posed three information literacy challenges between collaborative sessions, each related to the youth scholars' summer research and solar futures narratives.

Michelle subsequently analyzed the artifacts and interactional discourse associated with the workshops to inform our interpretation of the workshop outcomes and test the aforementioned design conjectures. She conducted a content analysis using the constant comparative method (Hsieh & Shannon, 2005) to categorize information sources and identify themes within and across the challenge responses and discursive contributions. By describing each workshop's artifactual and discursive outcomes, we hope to spark imaginative educative directions for supporting youth climate change education supporting youth to inform, imagine, and create civic renewal.

Workshop 1. Informing Renewal: What We Learn When We Learn to Search

An emphasis on utopian creative media does not preclude the benefits of using traditional information search strategies to find academic scholarly answers to particular questions. This is where we started with the youth participating in our program. Workshop 1, *Informing Renewal*, focused on what we learn when we learn to search. It began with a 20-minute video introduction by the workshop facilitator. [NOTE: the video can be found at *youtu.be/Oktd_cz4kQc*.]

Catherine first introduced herself as a librarian and then defined *information literacy* as follows:

> The set of integrated abilities encompassing the reflective discovery of information, the understanding of how information is produced and valued, and the use of information in creating new knowledge and participating ethically in communities of learning.

The workshop host further defined information literacy as seeking and obtaining knowledge and then reframing it through synthesis. She explained the processes involved in conducting a traditional digital information literacy search: exploring the internet to find, store, understand, share, and create new information. Finally, situating the youth with agency and responsibility related to the program's focus on energy transition and solar energy research, she asked the youth to consider all the misinformation they have seen concerning climate change. She explained that it is the job of scientists to decrease the spread of misinformation and increase the spread of high-quality information that can be used to impact climate change positively.

Finally, Catherine presented the *Informing Renewal Challenge*, inviting the youth to find information about issues of importance to the renewable energy work they were embarking on in the larger program, a research project integrating agriculture with photovoltaic electricity. They were asked to generate a list of keywords and then conduct an exploratory search in Google based on two questions:

- How might solar energy be used in conjunction with farming?
- What are some emerging trends in solar energy technology?

Following their viewing of the introductory video, the youth independently conducted an exploratory search, curating their favorite links along with a short description of their findings in a collective Google Doc.

In analyzing the youths' contributions to the *Informing Renewal Challenge*, it is first noticeable that the entire set of resources consisted of online informational articles. The majority of these sources were websites by (a) advocacy groups committed to sustainable and socially just energy transitions, (b) trade publications informing industry members of up-to-date photovoltaic technological innovations, and (c) energy providers offering articles for consumers. One youth explored articles published by the Pew Research Center, a nonpartisan fact tank.

Also of note is that all the resources shared were text-based, relying heavily on the written mode of communication and none of the resources explicitly included artistic elements. Moreover, in sharing their search results, the youth were all hyper-focused on the textual content; none referenced photographs, graphs, or other multimodal elements that co-comprised their selected informational resources.

The youths' written descriptions aligned with their orientation to reporting on facts rather than speculation. They offered either high-level summaries of information from multiple sources (e.g., "These have good information with agrivoltaics and how solar energy is beneficial to farms"), or specific information nuggets garnered from a single source (e.g., "Farming counts for 85% of global water consumption...solar panels produce 10% more power when plants are grown under them"). The trend of fact-focused reporting continued as the group gathered to share the results of their searches. Carlos's enthusiastic summary of information from his exploration typifies the youths' pattern of identifying emerging technological trends on the cusp of implementation on the immediate near-future horizon:

> ...solar energy is rising in demand and people are trying to implement it into many things, so like, obviously, wanting to try to implement it into cars. And apparently, they're going to put it into roads, and they're trying to develop solar powered paint, which is quite strange... they're making solar powered glass, which is also interesting.

Following these initial share-outs, Catherine invited the youth to reflect on their search process:

- Why did you decide to select these resources?
- Do you think that you found "good" resources?
- What rationales did you use to make those decisions?

All the youth evaluated the resources they selected as "good" either because of the novelty of information garnered (e.g., "it was interesting"), its relevance to their immediate shared solar research efforts (e.g., "I chose this because it was directly related to how solar energy can be used in farming"), or the benefit they perceived for their community (e.g., "more and more homeowners are buying solar panels—which is really good, especially in our area...since it's really sunny. Not only is it sustainable but people, in the long run, save money").

Using these initial responses as a catalyst, Catherine then defined three common criteria for determining how information is evaluated from a literacy perspective. These criteria include whether information is relevant to research needs (contextual in time and space), produced by an authoritative source (derived from reputable sources; how have markers of "authority" changed over time?), and reflective of societal values (information that doesn't "speak to us" doesn't get used; what are our current methods of valuation?). The element of *value* would move to the center during Workshops 2 and 3 when engagement with utopian creative media became the focus of collective engagement.

Workshop 2. Imagining Renewal: What We Learn When We Learn from the Arts

Workshop 2, *Imagining Renewal*, invited the youth to consider the roles utopian creative media play in conveying climate change information. The group collectively explored diverse examples of storytelling as a vital medium for conceptualizing utopia, including images, videos, music, and spoken word. They looked for context clues about how the future is imagined by the person(s) who created the media object and reflected on how each one made them think about our shared future: how it will be changed, can be changed, and must be changed. Finally, Catherine presented the *Imagining Renewal Challenge*:

- Engage with the selection of curated utopian creative media resources in the climate futures bibliography; identify 2 that inspire YOU and spark your imagination about the future.
- Search online to find a new utopian creative media resource of your own, one that moves you.
- Share your reflections, along with quotations, images, videos, links, etc., from each of your selected resources in the Google Slide Deck to inspire your team's imagination.

The youth were given a bibliographic list of utopian climate-focused creative media forms. [A version of this list can be found at *t.ly/pc2S*.] They were invited to explore and respond to the material independently, indicating how their selected media informed their utopian visions of climate change and provided a new context for understanding an environmentally sustainable future. After completing the challenge, the group met again to share their multimodal responses.

Altogether, the youth curated and commented on a wide range of media objects, exhibiting 12 slides with text, graphics, and weblinks to a wealth of resources: four song videos, three full-length movies, one short film, one climate fiction novel, one podcast, a video of a fashion show, and a solar punk contest announcement. Looking across these learner-generated artifacts and the subsequent discussion, we interpreted the youths' reflections as clustering under four categories:

- coming to new realizations
- setting intentions to take up new ethical responsibilities
- making predictions about the future
- imagining a future in which agency is possible

Some of the slides illustrate how youths' textual encounters supported their coming to new realizations about climate change and possible futures. One such example is shown in Victoria's engagement with a video of the *Balenciaga Fall 2020 Fashion Show, youtube.com/watch?v=9QMugKSXFWQ&t=1s*. Art director Nicke Bildstein-Zaar and creative director Demna Gvasalia, leaders in the global sustainable fashion movement, created a powerful visual post-apocalyptic commentary. In Victoria's words, they "Flooded the stage and had LED graphics on the ceiling mimicking thunder, fire and rain to make a statement on the climate change crisis." Having pasted a screenshot of a model walking the deluged runway wearing toe shoes, Victoria described the presentations' purpose as she saw it: "[They] had toe shoes in their collection to make a statement on how our fashion will have to evolve to live in a society where climate change gets worse." While sharing this artifact, Victoria expressed several new realizations:

> It shows us all again how many different modes of information we can sort of communicate with and think through in order to better understand… how the future is going to look… Fashion is going to make its way into the future with us…How do we think about ways to make [fashion] more sustainable and make it resemble the world we want to be in?

Victoria first communicated her expanded awareness on two fronts. First, her engagement with this media object allowed her to recognize new options for information sources that can be brought to bear on communicating about climate change and imagining possible and probable futures. Further, although the Balenciaga show was post-apocalyptic in its orientation, its portrayal of human resilience in the face of climate change nonetheless invited Victoria into a utopian imaginary space. Being in that space challenged her to new considerations of how everyday practices (decisions about practical footwear) could be adapted to a world of rising sea levels.

Beyond new realizations, the youths' interaction with utopian media objects spurred their setting new intentions in relation to climate change action. For instance, as a result of her encounter with Jeff Bandermeer's (2017) ambiguous utopian novel, *Borne*, Sora reflected:

> Borne presents a dystopian world destroyed by horrific biotech creatures, which presents the idea of how the exploration of science without considering its consequences on society can lead to dangerous effects…The book also helped inspire me to become more careful in designing things for technology and considering whether or not it would actually be useful for the community—or is it something I'm only interested in.

Exploring the far edge consequences of a societal lack of caution in technology development, Sora reflected on the potential for innovations to run amuck.

A serious scholar with a long-term goal of becoming a solar energy engineer, Sora came away from this encounter with a new intention to take up ethical responsibilities in her work as a community engineer. Rather than orienting to the dangers in a generalized way, she allowed this utopian text to shape her technological trajectory, explaining to her peers,

> You have to always look at whatever you're creating when it comes to technology. How is it going to affect society? How is it going to affect people — even if I don't intentionally want to affect them? Because every action that we create will always have a reaction, and I think that's why we need to be considerate with it.

Thus, Sora oriented to this ambiguous utopian text as a cautionary tale for her imagined future self that invited her and her peers into co-creative action.

Beyond expressing greater awareness, several youths made assertive predictions in their responses to the utopian creative media they explored. We saw these predictions as agentive insomuch as they forwarded and elaborated on counter-narratives to the prevailing status quo. Max, for instance, responded thus to the song and video, *Amazonia*, by Gojira, a French heavy metal band whose members are also activists in environmental, Indigenous, and animal rights:

> The song is really about how modern civilization is seen as the greatest thing and it is failing and that Indigenous tribes will be the ones to survive. Their ways of life have been refined over centuries, and they work—they allow people to survive. So, it's…bringing a highlight to that their ways of life have been refined to work with the earth, whereas our civilization now kind of contradicts or works against the earth. While modern civilization will eventually fall, indigenous tribes will remain as the peoples of the earth.

Max's prediction that Indigenous peoples will survive the coming effects of climate change is predicated on their understanding of *Amazonia*'s contrast of modern civilization's sustainable lifestyle with Indigenous ways of life. Their interaction with this musical text and the contrasting social and cultural practices it portrayed invited Max into utopian thinking that subsequently shaped their description of a better sociopolitical place not yet present when imagining their utopian solar future narrative later in the program. Using the medium of a year 2060 newspaper report, Max envisioned two neighboring trade partners, one rural and one urban. Both communities rely solely on renewable energy generation and "blend their tech seamlessly with the environment and focus on the materials and resources they are using."

Finally, we saw several of the youths' responses to the media objects they explored went beyond merely imagining the future, to envision a future with expanded opportunities for agency. For example, Samantha used her Instagram

Solarpunk Contest called Future Fiction

- Concept art for a future with floating villages
  - ○ Artwork by Commando Jugendstil
- We can imagine a future from this image based on the details of this work
  - ○ How to live in water because water levels are rising
  - ○ How to gather energy in the ocean effectively
  - ○ How we can implement more plants and greenery into our homes. (in this case village)

FIGURE 2.1 Samantha's Response to Floating Futures

account to search for information on solarpunk, a movement of utopian speculative fiction and art that imagines low- and high-tech uses of renewable energy. It also generates forms of social organization that place value on community and draws on utopian themes to offer hopeful futures perspectives in the face of climate change (Norton-Kertson, 2021). Samantha's slide exhibited a solarpunk artwork by *Commando Jugenstil*, a collective of artists, researchers, scientists, and activists (*solarpunk.it/*). Her commentary on the piece (Figure 2.1) can be interpreted as her envisioning of a future in which agentive responses to climate change are possible.

Here, Samantha extracts three dimensions of climate change response (water, energy, food) from her encounter with this information resource, exhibiting her confidence in people's ability to rise to the challenges of meeting their everyday needs in the face of changing geographical conditions.

Later, in sharing her slide, Samantha expressed her enthusiasm for the socio-interactional commitments of the community in which the artwork was created. Asked why she found the solarpunk fiction contest intriguing, Samantha's reply indicated an additional opportunity for agency: organizing future communities for co-action:

> I use Instagram a lot and I was curious, what I could find hashtags with that?…I found a solarpunk competition that happened, and they had a bunch of posts about artworks that people have submitted with different concepts…I found it interesting because…I hadn't seen that before, like, so many different [sections] to one contest, so we have like a bunch of winners.

In articulating her admiration for the breadth of concepts and "winners," Samantha is noting important aspects of solarpunk as a utopian genre. The way the solarpunk community operates offers a powerful narrative about the interactional capacities that might be beneficial in communication practices related to climate change. For Samantha, solarpunk is compelling because its authors and followers are interested in thinking about how to make a better future. It is optimistic and open; it makes room for a wide range of media objects by a wide range of people. These orientations are reflected in a solarpunk fiction contest where the goal is not to pick the best artist, rather to hold up ideas for hopeful futures and make sure all contributions get an airing so more people can interact with and be inspired by them.

Summary

Altogether, the youths' responses to the *Imagining Renewal Challenge* illustrate how interacting with utopian creative media acted as a vehicle for our vision of a pedagogical space that invites learners into the agentive, hopeful imagining of a better as yet unrealized civic-future. Moreover, linking these responses back to the *Informing Renewal Challenge* showcases the interplay of affordances and constraints of different types of information resources. The youth oriented to the scholarly information search as an opportunity to find out how people are tinkering with current energy systems. The information they gravitated to was focused on technological innovations at the horizon of integration with existing social systems. In contrast, using utopian media as an information literacy intervention established an opening to discuss civic and social [structural change] that needs to occur for climate collapse to be resolved. Utopias focused thinking on how to radically change those systems, destroy them, or create entirely new ones that answer the question of climate change. This is necessary to help youth consider possibilities for a future which cannot be imagined or accessed within the confines of technocratic intervention. Technologies that support climate intervention have a role, but remain tools rather than solutions in and of themselves.

Workshop 3. Creating Renewal: What We Make When We Learn from the Arts

Workshop 3, *Creating Renewal*, raised questions about how information can be used to make things—particularly in regard to issues of power and social value. The group reexamined evaluations of information's relevance, authority, and especially social value, in order to prepare the youth for the *Creating Renewal Challenge*: writing their own utopian-esque solar futures narratives. The discussion elaborated on questions initiated in Workshop 1: When we look for information, how do we decide what information to keep, what to discard, and what to create?

The workshop began with a spoken reenactment of a scene from Ursula K Le Guin's (1984) ambiguous utopian novel, *Always Coming Home*. Le Guin introduced the audience to a far-future state, a matriarchal, post-capitalist, and post-industrial society that conveys renewal as central to societal well-being. The selection read is from a dialogue in which an archivist responds to a question about determining the value of information in relation to storage and retrieval: How do you decide what to keep and what to throw away?

> The Archivist: "Keeping grows; giving flows." Giving involves a good deal of discrimination; as a business it requires a more disciplined intelligence than keeping, perhaps…Books no one reads go; books people read go after a while. But they all go. Books are mortal. They die. **A book is an act; it takes place in time, not just in space. It is not information, but relation.**

Invited to comment on the quote above, Sora reflected:

> I think that we hold more value to the things that are limited, and if they are unlimited, then they lose value. I think that's what [s]he meant by discrimination because, well, if we keep some stuff and we lose some stuff then we're choosing which one's more valuable than the other. It's kind of like life. I mean, we choose. And why it's immortal is because we as humans tend to enjoy life more because there is an endpoint… that's why [s]he…discards the books, so people can enjoy it more for the time that it still exists.

Information literacy involves understanding how information is collected, protected, and made available, but also how information is never value-neutral in that it entails action. As such, learning how to discriminate between different types of information is often about realizing that we have a choice in the type of information we use and, further, choosing how best to use it. As people, we choose what we value, and thereby we choose what we keep. We ultimately choose what we can't value, or we don't have time or the energy or the ability to process—and this matters for thinking about the future. It matters to consider how we move forward as a species and on a planet that is in danger. It matters how we see utopia.

If we create certain types of scholarly texts, if we create certain types of artistic media, we generate certain types of ideas. The idea of venerating certain ideas is a way of showing respect and honor, and this shapes how we view utopia. As an example to think with, the group collectively viewed *The seeds: ReGeneration t.ly/_Z0H*, a short film that celebrates choreographer Rulan Tangen's collaboration with tribal leaders and dancers across the US Four Corners region. Her work combines community visioning exercises with dance performances centering Indigenous artistic and ecological knowledge to confront ecological collapse.

Situating the video is an example of engaging with climate change through art, Catherine invited the youth to reflect on how it challenges what we think of as information and ways of knowing when it comes to living alongside nature. The group discussed questions of authority: who gets to socially value different forms of information, particularly regarding information from lived experience, which can be reduced through socially totalizing structures, and Indigenous knowledge and Indigenous ways of understanding, which have often been delegitimized. Extending the conversation, Catherine asked if and why the video sparked the youths' own imagination and creativity: how do we use information to decide what to create? How can we use information to tell a story?

Creating Renewal Challenge: Connecting Information to Experience through Storytelling

One way to decide the value of different media objects is to use the information that allows us to best tell a story. When people engage in storytelling, they take in information they have received and rearrange it before putting it back into the world. All stories in all media forms have many moving pieces that correspond with politics, economics, culture, religion, science, and ecology. Good stories of energy transition futures, for instance, are not just set pieces to describe sustainable transitions to solar power or wind energy. They are entire worlds.

Thus, our workshops were designed to culminate in youths' speculative fiction writing. With the workshop ideas and their experiences with both scholarly articles and utopian media in mind, youth were challenged to produce their own utopian media creations. The third author, Steve, led the youth in a series of collaborative discussions and co-constructive activities to story possible future energy transitions in their local community. He put to work the multiplicity of intentional information resources students drew on during their workshops. Each session included cohort discussions and either small group collaboration or independent inquiry and authoring.

The sessions progressively developed individual and collaborative writing that challenged students to articulate utopian-oriented stories about preferred energy futures. Sessions also involved a recurring process of feedback in order to loop insights from peers into iterative refinements into story drafts to fuel this progression. It also enlisted insights from collaborative feedback into subsequent discussion and ongoing writing. Futures thinking thus challenged youth to imagine how societies might utilize solar technologies to transition into a post-carbon energy system, how their own agentive actions might contribute to this transition, and ultimately, how possible transitions realized through this socio-technical interplay might positively impact planetary health.

Discussion

Climate change is a global phenomenon that occurs on geographic and temporal scales nearly impossible to conceptualize fully. It is the culmination of numerous interlocked systems that build on one another and coalesce in ways that make it difficult to imagine how they can be changed or undone. Transportation, fuel extraction, agriculture, deforestation, industrial development, mass migration, mining: these and other large-scale systems contribute to climate change, and to resolve climate collapse, they must be altered in some way.

Due to the complexity of thinking through these necessary changes, we believe that futures thinkers must be provided with alternative means for developing tactics. Doing so entails using creative media that constitutes new ways of knowing, which work beyond the boundaries inherent to traditional scholarship. Additionally, we call for teachers to present learners with a wide range of creative media that conveys utopian visions. Further, we encourage teachers to deploy utopian media that wrestles with some element of climate change or with the alteration/destruction of systems that contribute to global climate collapse. Such media should also provide learners with opportunities to imagine climate futures that are more egalitarian and just. Finally, while dystopian subgenres have a place in the learning environment, they ought to be supplemental to explicit utopias and/or ambiguous utopias that evoke hope and thereby agency and action.

Information literacy innovations can be designed to help students conceptualize a utopian future through immersion in creative media, specifically media that undertakes a utopian project. Thus, in combination with critical assessment of utopian texts, learners should be encouraged to apply scholarly research and futures thinking to develop possible solutions to climate problems by coupling reasoning and imagination. Teachers might, for instance, task learners to work collectively to imagine and develop an environmentally liberatory, utopian world. Multiple creative media forms can serve as a platform for youths' creativity, including, for instance, photography (Trott, 2020), film (Littrell et al., 2020), multimedia narratives (Jordan et al., 2021), and musical compositions (Tobias et al., this volume). While such authoring often elicits stories that explore the potential negative consequences of climate change, it also creates opportunities for youth to resist such outcomes and imagine desirable alternatives for themselves and their communities.

Utopian thinking as a progressive movement is fueled by individual's values and dreams, as authors and creators, and also by collective efforts illuminated through information literacy, as citizens and co-inhabitants of a planet in crisis. There is a difference between utopian thinking with and without information literacy, between the futures we might envision alone and those we might envision by exploring the visions of others. Centering utopia in climate change education steeps learning in agentive action rather than limiting youth to learning about climate change. It seeds and sustains hope rather than promoting pessimism, is future-oriented rather than limited to what currently exists, and provides opportunities to seek, critique, produce, and share informative media objects that inform, imagine, and create renewal.

References

Angheloiu, C., Sheldrick, L., & Tennant, M. (2020). Future tense: Exploring dissonance in young people's images of the future through design futures methods. *Futures*, *117*, 102527. doi:10.1016/j.futures.2020.10252

Atwood, M. (2003). *Oryx and crake*. First Anchor Books.

Bai, X., van der Leeuw, S., O'Brien, K., Berkhout, F., Biermann, F., Brondizio, E. S., Cudennec, C., Dearing, J., Duraiappah, A., Glaser, M., Revkin, A., Steffen, W., & Syvitski, J. (2016). Plausible and desirable futures in the Anthropocene: A new research agenda. *Global Environmental Change*, *39*, 351–362. doi:10.1016/j.gloenvcha.2015.09.017

Bandermeer, J. (2017). *Borne*. Farrar, Strauss, and Giroux.

Beach, R., Boggs, G., Castek, J., Damico, J., Pano, A., Spellman, R., & Wilson, N. (2020). Fostering preservice and in-service ELA teachers' digital practices for addressing climate change. *Contemporary Issues in Technology and Teacher Education*, *20*(1), 4–36. https://www.learntechlib.org/primary/p/210433/

Bellamy, E. (1888). *Looking backward, 2000–1887*. Ticknor and Co.

Butler, O. (1993). *Parable of the sower*. Grand Central Publishing.

Cattell, J. (2021). "Change is coming": Imagined futures, optimism and pessimism among youth climate protesters. *Canadian Journal of Family and Youth/Le Journal Canadien de Famille et de La Jeunesse*, *13*(1), 1–17. doi:10.29173/cjfy29598

Coleman, J., & Pankl, L. (2020). Rethinking the neoliberal university: Critical library pedagogy in an age of transition. *Communications in Information Literacy*, *14*(1), 66–74. https://pdxscholar.library.pdx.edu/comminfolit/vol14/iss1/5

Coogler, R. (2018). *Black panther*. Walt Disney Studios Motion Pictures.

Crane, T. (2001). Intentional objects. *Ratio*, *14*, 336–349. http://www.timcrane.com/uploads/2/5/2/4/25243881/intentional_objects.pdf

Davidson, J. P. L. (2019). My utopia is your utopia? William Morris, utopian theory and the claims of the past. *Thesis Eleven*, *152*(1), 87–101. doi:10.1177/0725513619852684

Epstein-HaLevi, D. Y., Silveira, F., & Hoffmann, M. (2021). Eco-activists and the utopian project: The power of critical consciousness and a new eco-imaginary. *International Studies in Sociology of Education*, *30*(1–2), 13–33. doi:10.1080/09620214.2020.1864223

Finn, E., & Wylie, R. (2021). Collaborative imagination: A methodological approach. *Futures*, *132*, 1–11. doi:10.1016/j.futures.2021.102788

Freire, P. (2021). *Pedagogy of Hope: Reliving pedagogy of the oppressed.* Bloomsbury Academic.

Gidley, J. (2017). *The future: A very short introduction.* Oxford University Press.

Gilman, C. P. (1915/1998). *Herland.* Dover Publications.

Giroux, H. A. (2007). Utopian thinking in dangerous times: Critical pedagogy and the project of educated hope. In M. Cote (Ed.), *Utopian pedagogy* (pp. 25–42). University of Toronto Press. doi:10.3138/9781442685093-004

Herobeat Studios (Developer). (2022, July 19). *Endling: Extinction is forever.* HandyGames.

Hicks, D. (2014). *Educating for hope in troubled times: Climate change and the transition to a post-carbon future.* Trentham Books Limited.

Holfelder, A. K. (2019). Towards a sustainable future with education? *Sustainability Science, 14*(4), 943–952. doi:10.1007/s11625-019-00682-z

Hsieh, H. F., & Shannon, S. E. (2005). Three approaches to qualitative content analysis. *Qualitative Health Research, 15*(9), 1277–1288.

Jasanoff, S., & Kim, S. (2015). *Dreamscapes of modernity: Sociotechnical imaginaries and the fabrication of power.* University of Chicago Press.

Jenkins, H., Peters-Lazaro, G., & Shresthova, S. (2020). *Popular culture and the civic imagination: Case studies of creative social change.* New York University Press.

Jordan, M. E., Zuiker, S., & Bernier, J. (2021). The future is open and shapeable: Using solar speculative fiction to foster learner agency. *Literacy Research: Theory, Method, and Practice, 708*(1). doi:10.1177/23813377211028263

Kafai, Y. B., & Peppler, K. A. (2011). Youth, technology, and DIY: Developing participatory competencies in creative media production. *Review of Research in Education, 35*(1), 89–119. doi:10.3102/0091732X10383211

Le Guin, U. K. (1984). *Always coming home.* University of California Press.

Levitas, R. (1990). *The concept of utopia* (1st ed.). Syracuse University Press.

Levitas, R. (2013). *Utopia as method: The imaginary reconstruction of society.* Palgrave Macmillan.

Littrell, M. K., Tayne, K., Okochi, C., Leckey, E., Gold, A. U., & Lynds, S. (2020). Student perspectives on climate change through place-based filmmaking. *Environmental Education Research, 26*(4), 594–610. doi:10.1080/13504622.2020.1736516

McKnight, H. (2020). 'The oceans are rising and so are we': Exploring utopian discourses in the school strike for climate movement. *Brief Encounters, 4*(1). doi:10.24134/be.v4i1.217

Miller, C. A. (2022). Redesigning political economy: The promise and peril of a green new deal for energy. In B. Fong & C. Calhoun (Eds.), *The green new deal and the future of work* (pp. 270–294). Columbia University Press.

Miyazaki, H. (2012). *Nausicaä of the valley of the wind, Deluxe Edition I.* VIZ Media LLC.

Moser, S. C. (2007). More bad news: The risk of neglecting emotional responses to climate change information. In S. C. Moser & L. Dilling (Eds.), *Creating a climate for change: Communicating climate change and facilitating social change* (pp. 64–80). Cambridge University Press.

Norton-Kertson, J. (2021, August 31). *What is solarpunk?* https://justinenortonkertson. medium.com/submissions-for-solarpunk-anthology-e4015346c1de

Ojala, M. (2017). Hope and anticipation in education for a sustainable future. *Futures, 94*, 76–84.

Pasqualetti, M. J. (2021). *The thread of energy.* Oxford University Press.

Pendleton-Jullian, A. M., & Brown, J. S. (2018). *Design unbound: Designing for emergence in a white water world* (Vols. 1 & 2). The MIT Press.

Robinson, K. S. (2016). Remarks on utopia in the age of climate change. *Utopian Studies, 27*(1), 1–15.

Robinson, T. (2020, October 20). We asked Kim Stanley Robinson: Can science fiction save us? *Polygon.* https://www.polygon.com/2020/10/20/21525509/kim-stanley-robinson-interview-science-fiction-utopias

Roddenberry, G. (Executive Producer). (1966–1969). *Star Trek: The original series* [TV series]. Desilu Productions (1966–67) and Paramount Television (1968–69).

Sabry, S. (2021). Traversing utopian nature: A proto-ecological approach to Thomas More's utopia. *Journal of Scientific Research in Arts (Language & Literature), 3,* 160–173. doi:10.21608/jssa.2021.73381.1257

Salemink, O., Bregnbæk, S., & Hirslund, D. V. (2020). *Utopian movements, enactments and subjectivities among youth in the Global South: Ethnographic perspectives.* Taylor & Francis.

Sargent, L. T. (1994). The three faces of utopianism revisited. *Utopian Studies, 5*(1), 1–37. http://www.jstor.org/stable/20719246

Seale, M. (2016). Enlightenment, neoliberalism, and information literacy. *Canadian Journal of Academic Librarianship, 1*(1), 80–91.

Siegner, A., & Stapert, N. (2020). Climate change education in the humanities classroom: A case study of the Lowell school curriculum pilot. *Environmental Education Research, 26*(4), 511–531. doi:10.1080/13504622.2019.1607258

Sunrise Movement. (2023). *We are the climate revolution.* Accessed March 28, 2023. https://www.sunrisemovement.org/?ms1/4SunriseMovement-WeAreTheClimate Revolution

Tobias, E., Bartlett, K., Jordan, M. E., & Zuiker, S. (this volume). Addressing climate change and sustainable energy futures through creative music engagement. In R. Beach & B. E. Smith (Eds.), *Youth created media on the climate crisis: Hear our voices* (pp. 146–165). Routledge.

Trott, C. D. (2020). Children's constructive climate change engagement: Empowering awareness, agency, and action. *Environmental Education Research, 26,* 532–554.

Trott, C. D. (2021). Youth-led climate change action: Examining multi-level effects on children, families, and communities. *Sustainability, 13*(22), 12355. doi:10.3390/su132212355

Turrentine, J. (2022, February 28). IPCC: We cannot look away—climate risks are cascading. *Natural Resources Defense Council.* https://www.nrdc.org/stories/ipcc-we-cannot-look-away-climate-risks-are-cascading

Walsh, E. M., & Cordero, E. (2019). Youth science expertise, environmental identity, and agency in climate action filmmaking. *Environmental Education Research, 25*(5), 656–677. doi:10.1080/13504622.2019.1569206

Wells, H. G. (1895/2002). *The time machine.* Readers Library Classics.

Wright, S., Han, V., & Hogan, C. (2021, March). The rise of eco-anxiety. *Force of Nature Report.* https://bit.ly/3jkUHvg

3

GENERAL ECOLOGY AND SPECULATIVE PEDAGOGIES

Youth Digital Media Practices for Climate Justice

David Rousell, Thilinika Wijesinghe, and Amy Cutter-Mackenzie-Knowles

Chapter Orientation

As digital media increasingly reshape what it means to sense, think, and learn in the twenty-first century, today's young people literally inhabit "a new world, a world characterized by a vastly expanded and deterritorialized sensorium" (Hansen, 2015, p. 161). This newly technologized world can be understood as a "general ecology" that directly embeds digital technologies within biological, social, and environmental systems (Hörl, 2017), including those associated with climate change. Digital media now modulate nearly every facet of young people's engagement with climate change (Boulianne et al., 2020), from the latest climatological modeling to the mass organization of international protest movements (Mayes & Holdsworth, 2020). Not only do young people gather and exchange knowledge about climate change through digital media, they also learn, socialize, play, and invent using digital technologies which are intricately entangled with climate change as a planetary event (de Freitas et al., 2020; Rousell et al., 2021).

This chapter explores how young people's digital media practices contribute to climate justice through the conceptual figuration of a general ecology. This figuration situates digital media technologies as elements of young people's natural environments so that we can analyze youth digital media practices as intricately interwoven with atmospheric and climatological processes. We suggest that such an approach is aligned with the current technologization of Earth-scale ecological systems. It also indicates wider socio-ecological transformations of planetary life associated with the Anthropocene epoch.

DOI: 10.4324/9781003335276-5

To illustrate this "general ecology" approach to youth digital media studies, we draw on two key examples. The first involves the co-development of Climate Action Adventure! (*ccme.app*), a web application (App) co-designed with young people across Australia that connects virtual worlds and characterizations with live social media platforms and climate data. The second example involves the co-production of speculative online drama performances with young people in Australia and Sri Lanka, focusing specifically on exploring alternative "post-Anthropocene" futures based on current climate data and possibilities of response.

We analyze the co-design of these two examples as specific configurations of climate change education and digital technologies. We identify three ways young people are currently mobilizing for climate justice through digital media: *anonymous participation, character-building,* and *world-building.* We show how each of these methods involves both a speculative and pedagogical investment in climate justice (Rousell et al., 2017) and suggests ways of integrating a "general ecology" approach into everyday teaching and learning environments.

Climate Justice, Affect, and Digital Life

The recent YouthStrike4Climate and SchoolStrike4Climate movements, along with Extinction Rebellion and numerous other youth climate activism groups, have brought unprecedented public attention to young people's political investments in climate justice. These youth-led movements are publicly reframing climate change as a profound issue of generational, social, and environmental injustice, emphasizing the failure of capitalist ideals and techno-scientific regimes in securing a sustainable future (Rousell & Cutter-Mackenzie-Knowles, 2023). In refuting public perceptions of youth as politically disaffected and disengaged, the strikes demonstrate young people's capacity to self-organize a dissident political action on a global scale. The climate protests, which launched in 2019, can be considered the first mass mobilization of an "Anthropocene generation," as children and young people refuse to comply with an education system that does not take ethical responsibility for a changing climate.

Digital media technologies have been central to the recent irruption of youth climate protests movement across the globe. Young people's digital practices have been widely researched and theorized in education (Ringrose, 2011) and youth studies (Caron et al., 2017; Fullagar et al., 2017). However, the political implications of digital media for young people's engagement with climate education and activism are only beginning to be researched and conceptualized (see, for instance, Boulianne et al., 2020; Maher & Earl, 2019; McLean & Fuller, 2016; Rousell et al., 2023).

Young people are now growing up within a vastly expanded milieu of digital media networks, artificial intelligences, and sensory technologies that intersect in complex ways with the onset of climate change and the general ecology of the Anthropocene (Cutter-Mackenzie & Rousell, 2018). Not only do young people gather and produce knowledge through digital media, but they also work, socialize, play, and modulate their affective states through digital platforms. Moreover, their consumption of these digital platforms changes variably and provides multiple new avenues for young people to generate and exchange experience, knowledge, sensation, and feeling (McGillivray et al., 2016).

Young people's diverse engagements with digital media can be understood as a radically expanded investment and ecological distribution of affect. As we use it in this chapter, the term affect describes the power to affect and be affected within dynamic confluences of relations and desire that could also be termed "atmospheric" (Manning, 2012, 2013; Massumi, 2002, 2015). Affects draw bodies together and/or apart in ways that are often unconscious and unintentional. They intersect as much with nonhuman bodies and events as those defined as human. This makes affect inherently ecological, as the specific unfolding of each affective relation alters the general potential for bodies to think and act in the world (Massumi, 2015; Protevi, 2013). Therefore, what we describe here as a general ecology includes a political investment of desire, vitality, and feeling into circulation within a complex ecology of felt relations between human and nonhuman bodies of many kinds.

(En)countering "Environmentality" through Digital Media

Erich Hörl's (2017) introduction to general ecology is compelling as a way of considering the force of relational counter-practices that young people are currently developing through the digital. Hörl argues that we currently find ourselves

> at a very specific point in the history of relationality that brings out the question and the problem of relationality much more radically than ever before: relational technologies and an algorithmic governmentality reduce, regulate, control, even capitalize relations to an enormous extent, and precisely in so doing, become essential to the form of power of *Environmentality*.
> (*p. 8*)

The concept of "environmentality" is crucial to grasp the implications of general ecology, both as a current state of political affairs and as a critical analytic figuration of the climate crisis. As an environmentally distributed form of power, environmentality thrives on the massive proliferation of digital technologies at a scale that transforms the ecological functioning and potentiality of the planet. This enables new forms of governance and social control to emerge, which are far more distributive and atmospheric than previous regimes of disciplinary order and centralized modes of governance. Environmentality is, in this respect, not only located within human communication networks but even more widely distributed through the mass deployment of internet-enabled sensors and actuators that increasingly regulate and intervene in environmental processes (Gabrys, 2020). Moreover, this technical proliferation of environmentality makes new forms of "exploitation of relations possible" (Hörl, 2017, p. 8) through the algorithmic mining and reduction of relations to calculable data points, which can be manipulated through artificial intelligence to serve capitalist interests.

Young people's use of digital media technologies to stage mass climate protests can, in this sense, be seen as exploitations of advanced capitalist environmentality to generate new and different ecologies which are resistant to the profit-driven agendas of governments and corporations. To the extent that general ecology implies a complete ecologization of planetary life, thinking, sociality, technics, and governance, the critical question is *what ecologies* can effectively sustain human and nonhuman lives into the future. As Hörl (2017) notes, there are literally thousands of ecologies operating across the Earth today:

> ...ecologies of sensation, perception, cognition, desire, attention, power, values, information, participation, media, the mind, relations, practices, behavior, belonging, the social, the political—to name only a selection of possible examples. There seems to be hardly any area that cannot be considered the object of an ecology and thus open to an ecological reformulation.
>
> *(p. 1)*

While these ecologies proliferate continuously, the ecological connectivity that *everything* has with one another contributes to the peril or the growth of each relationship within the ecology. For example, just like the food cycle that is interconnected and interdependent, the ecological web between biological, social, digital, and environmental systems is also connected. It forms newly complex relationships between entities across different spatial scales and temporal durations. A primary way young people become a part of this meshwork is by intermingling with the digital ecology, making them part of a larger bioorganic and geomorphic matter ecology. The digital ecology not only enables young people to express their views on various issues, but also capacitates them to live with and through complex geo-bio-social *problems* such as climate change, building new forms of value and sociality while exploring and re-exploring the problem from an ever-changing series of relational perspectives.

We believe that it is essential that ecologies remain open to relational differences rather than closed, and therefore simply repeating the same ecologies. This openness to difference is essential if ecologies (including digital ones) are going to produce new forms of value that deviate from and exceed capitalist profit, including valuing biodiversity and climatological stability beyond mere resourcing and instrumental use-value. As Erin Manning (2023) writes, "experience is the eruption of difference that necessitates new forms of valuation at every turn... every shift alters the whole, reorienting it in the interplay of existence, thereby shifting all valuations of what counts" (pp. 48–49). This variation highlights that different ecologies of experience among young people contribute to the formation of new knowledge and harnesses emerging forms of relationality and value.

These ecologies of experience are distinctly situated and continuously evolving within historical plateaus that inform the collective social movements of young people by imbricating them within the meshwork of a general ecological situation. The relationship between young people and technology rearranges and constantly keeps evolving within this general ecological web. Hence, the forms of value and subjectivity are always changing. This ecological interaction between young people and technology generates new responses to the issue of climate change while remaining part of the same messy, entangled world that is generating the catastrophe, effectively collapsing the possibility of a distanced, neutral, or "god's eye view" of the climate crisis.

Affect, Ecology, and the Digital

Young people's capacity to affectively connect dispersed and previously unreachable communities makes the digital a politically powerful ecological platform for young climate activists. Yet, it is also subject to radically expanded regimes of capitalist surveillance and predatory monetization (Taylor & Rooney, 2016). As de Freitas et al. (2020) note, young people's immersion in digital environments makes them "increasingly sensitive to affect through these systems... young people today labour in the affect economy, their biopower tapped and circulated through social networks" (p. 25). As we have argued previously, what is politically at stake in such developments is not simply the regulation and utilization of digital media networks to distribute information. It is also the digital environmental governance and redistribution of experience as an atmospheric modulation of young people's political engagement with climate change (Rousell et al., 2023).

Digital media are critical sites through which the new power regimes of environmentality are both exercised and challenged. This directly impacts young people's everyday life experiences. Specifically, young people's affective investments in digital media can variously ease, amplify, express, and dramatize the existential anxiety that accompanies life in an age of climate change (Verlie, 2019). And yet, as de Freitas et al. (2020) argue, the challenge is how to "flip that feeling" of existential anxiety and hopelessness so that the stakes of ecological relationality "are felt not only as here and now, and for us, but for the 'total environment' of an Earth and cosmos, in all its diversity" (p. 26, emphasis in original). This makes the digital an increasingly urgent context for young people to design and develop educational interfaces and platforms which are effectively sensitized and responsive to the problem of climate change.

Young people's environmental behavior is vastly influenced by the digital technological impressions they have and mingle with at every single moment. The figuration of a "general ecology" helps us become sensitive to the constantly weaving political and theoretical elements that contribute to the conceptual relations formed in and through young people's connections with digital technologies. In a wider sense, this is what affects young people's decision-making to connect with climate protests and express their political thinking. Such thinking is not only "human" because young people continuously live in an ecology of multiplicities and collaborations between human and nonhuman agents. This generative space creates "new thinking of togetherness and of great cooperation of entities and forces" (Hörl, 2017, p. 4), such that nonhuman forces, elements, and entities also become enlisted in the formation of social movements through the digital. Growing up within expanded digital media networks has enabled young people to become familiar with digital technologies as fluid and agentic interfaces that connect directly to changes and modulations in their environments. This has become evident in the complex formation of the youth climate protests over the past three years. As a result, youth climate activists think and interact in digital media environments that extend beyond the scope of a human-centered subjectivity to connect with more-than-human informatic flows and environmental milieus.

Speculative Pedagogies of Fabulation through Digital Media

Our previous research has shown that young people engage with digital media ecologies in ways that far exceed instrumental use-value and the desire to control or impact their environments to achieve predetermined aims (Rousell & Cutter-Mackenzie-Knowles, 2022; Rousell et al., 2017; Rousell et al., 2021). We have found speculative fabulation to be a critical aspect of youth digital media practices, which hold significant transformative potential for responding to problems as complex as climate change.

In our recent work, we have drawn on Deleuze's (1967) concept of "dramatization." We activate this concept to theorize young people's use of speculative fabulation to stage creative modes of response that shift the terms of value through which climate change is encountered. Deleuze uses the notion of dramatization to describe how ecological systems evolve in response to the problematic conditions of their environments while showing how these developmental processes always incorporate speculative elements into their very constitution. We have summarized this previously:

> Deleuze's "dramas" are processes that are constantly in motion—dynamic, and always occurring in response to the conditions of a problem which defines their coming into being. In a famous example, Deleuze describes the evolution of the human eye as a response to the problem of light under varying conditions. Visual perception is staged as an ongoing dramatisation of this very problem, where virtual potentials to sense and "see" are actualized through the act of perception within a dynamic ecology of differential relations. Deleuze refers to these dynamic ecologies as "spatio-temporal dynamisms" that outline dramatization operations across multiple series and powers of events.
>
> *(Rousell et al., 2023, p. 6)*

Importantly, this suggests that all developmental processes and activity involve a continuous production of subjectivity which arises from the context or milieu of an ecological problem rather than being bounded or concentrated within the human subject (Rousell, 2019). We are attracted to the possibility of theorizing youth digital media practices as ways of speculatively dramatizing the problem of climate change. This dramatization acts like a surface (or stage) through which new forms of identity and sociality can be constructed. Young people can then participate in the speculative construction of new forms of subjectivity that populate the intensive ecologies of digital worlds. These include anonymous forms of participation in online environments through which individuals remain unrecognizable and the creation of entirely new characterizations through the production of avatars, bots, and algorithmic entities. An example of this can be seen in Figure 3.1, which shows an avatar character created by a 12-year-old co-researcher for the Climate Change and Me project (2014–2016), which we discuss in more detail in subsequent sections.

As Cull (2013) notes, Deleuzian dramatization does not mirror the world as it is. Rather, it forms a surface of fabulation for staging a plurality of worlds and identities through intensive becoming. Sholtz (2016) further explains that "the powers of the false, the simulacrum, fabulation, performativity and aesthetic artifice serve as Deleuze's palette for imagining the creative and dramatic nature of life" (p. 51). Ethical and ecologically responsive engagements with these speculative "powers of the false" have become increasingly urgent in an age of post-truth politics, conspiracy theories, climate change denial, and widespread distrust in science (de Freitas, 2020; de Freitas & Truman, 2021). In what follows, we show how speculative pedagogies of fabulation become crucial devices for young people engaging with and responding to climate

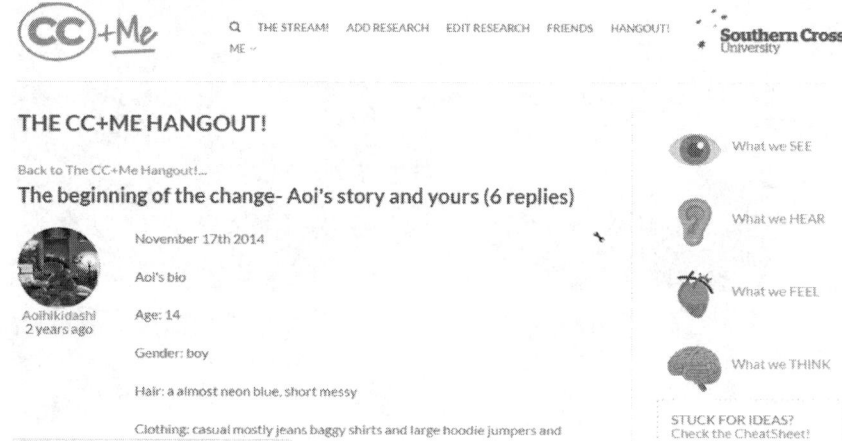

FIGURE 3.1   Avatar Character Created by a Young Person Using the Digital "Hangout" Space Originally Designed for the Climate Change and Me Project (2014–2016)

change. Moreover, these pedagogies have significant implications for cultivating young people's political engagement with environmental justice through digital media.

Climate Action Adventure!

In 2014, six years before the youth climate protests erupted worldwide, Cutter-Mackenzie-Knowles and Rousell began a project titled *Climate Change and Me* (CC + Me) in the Northern Rivers region of New South Wales, Australia. The project was a three-year collaboration with 135 children and young people (aged 9–14), investigating their experiences with climate change in their everyday lives. The project enabled young people to explore their own creative, aesthetic, political, and ethical responses to climate change through in-person workshops and various customized digital platforms. Young people then worked directly with the researchers to develop new modes of climate change education, including a transdisciplinary climate change curriculum that schools across Australia have since adopted. The project was highly generative, with a traveling exhibition that featured young people's art, research, filmmaking, and writing about climate change, much of which has also been published in previous work. Figure 3.2 shows a range of creative outputs from the project, including digital photographs, drawings, and a printed booklet of children's research which was distributed through schools, galleries, and libraries across Australia.

FIGURE 3.2 Examples of Creative Work by Young Co-researchers in the Climate Change and Me Project

Several years after the conclusion of the CC + Me project, the project team was funded to develop a climate change education App that could further extend the project's outcomes and reach new generations of children and young people. The CC + Me 2.0 project involved working with 20 young people from across Australia as co-designers to develop a climate education mobile application. The co-designers were invited through our partnerships with youth-led environmental activism groups such as Australia Youth Climate Coalition *aycc.org.au* and SEED Indigenous Youth Climate Network *seedmob.org.au*. As a result, children and young people were invited to co-design the Climate Action Adventure! Application *ccme.app* through a series of face-to-face and online workshops.

The young co-researchers of the CC + Me 2.0 app development project involved themselves in a creative, participatory, and applied co-design process that was collectively guided. As collective responses were noted, the workshops were mainly guided through speculative questions of ontology, epistemology, and ethics. Speculative visualization was used through an activity where young people were invited to imagine looking over a wall and to describe the kinds of futures they saw. The responses were often very different from one to another. For some, the future was pitch black, while others saw raging bushfires.

Drawing on our previous project, we reviewed the speculative digital practices developed by young people and invited the co-designers to test all other climate change mobile applications that were already available to understand their experiential design and functionality. The young people's responses were mainly focused on activism based on their personal experiences. Their discussions often revolved around social media platforms such as Twitter, which young people thought could be an effective platform to communicate short yet powerful messages across to the broader public to activate action on climate change.

Young people used the Marvel platform to collectively design what the app might look like and what it would include (see Figure 3.3). This took us closer to what is now often termed "experience design," and more specifically, to the design of "affective interfaces" (Fritsch, 2011). In these interfaces, the aim and outcome is not a product with predetermined functions and use-value but a field of experience that shifts with the affective investments of whoever (or whatever) is interacting with it (Brunner & Fritsch, 2011).

As the workshops continually evolved around the designing of the app, the general ecological nature was such that the co-designers, the young people, were using technology to relate experiences and to receive inspiration on how they could build and design this Climate Action Adventure! Application to combat climate change inaction. The process was continuously in flux, and, for each moment, the co-designers generated new ideas and concepts that collectively contributed to how the app eventually took form. The online workshops

FIGURE 3.3   The Climate Action Adventure! App Home Screen Where Characters Were Designed by the Co-researchers (*ccme.app*)

continuously facilitated discussion on what the application might include and how it might perform, which led to a series of speculative images which guided the design process (as discussed by Rousell et al., 2023):

- The feeling of being "suspended" in a kind of airship or dirigible while receiving continuous updates of climate data and concerns from around the world.
- Being able to "travel" to different parts of the world to encounter the specific impacts of climate change on particular creatures, communities, and environments.
- "Hanging out" in a digital learning environment or campus with a range of spaces such as a cinema, library, café, art gallery, and so on.
- The creation of a "mixed reality," inviting players to take on fictional personas, avatars, and adventures while maintaining a close connection with actual events related to climate change impacts, politics, activism, social media, and education initiatives.
- Being able to "take the pulse" of global youth climate activism movements through social media and hashtags while also launching new activist campaigns from the app itself.
- "Crowd sourcing" climate concerns from young people globally and linking to scientific data on climate impacts in specific ecologies and regions.
- Attaining some kind of "experience points" or other indication of movement toward addressing climate change within the world of the app.

At this stage, our design process was joined by Red Sky Media, a professional digital development agency that had worked with us on the original CC + Me project. Together with the app developer, Red Sky, we were able to map what feasible trajectories we could take to develop the app. Considered within the general ecology of things, young people's virtual ideations were now being actualized, collectively generated, and in continuous connection with digital technologies and the problem of climate change. The resulting app has been highly generative, enabling young people worldwide to contribute their concerns about climate change and activate their climate action initiatives (see Figures 3.4 and 3.5).

Speculative Drama and Youth Digital Practices of "Spaceulation"

Our second research example derives from Wijesinghe's (2022) PhD research study, which looked at how children and young people express their worldviews of the future on climate change, through an arts-based methodology called speculative drama. As we discussed, the new aesthetic methodology

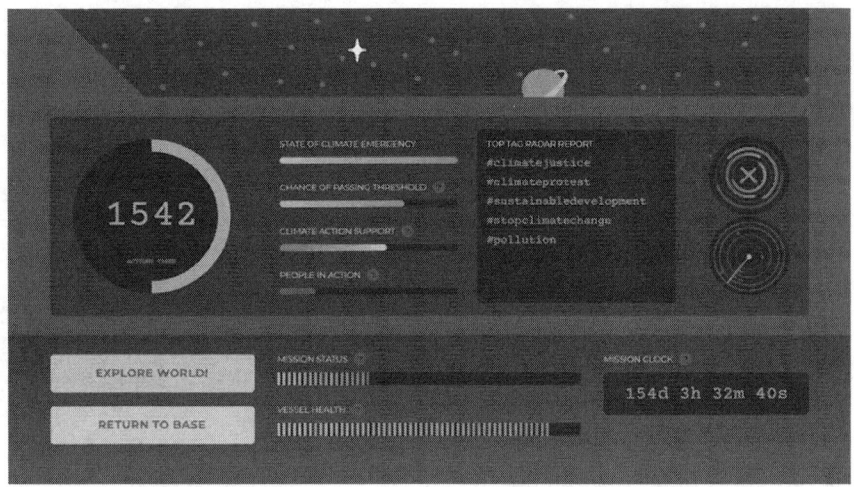

FIGURE 3.4   The Climate Action Adventure! Dashboard Provides a Global Overview of Climate Activism on Social Media, Climate Science, and
Youth-generated Activity Data

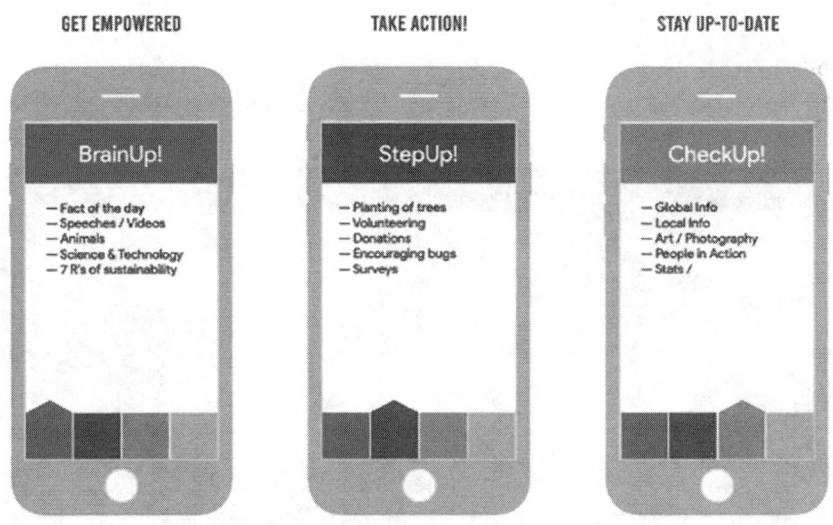

FIGURE 3.5   In the Lead up to the Development of the App, the App Developer
Mapped the Digital App Interfaces with the Suggestions of the Co-
researchers to Understand What the Final Web App Might Include

stemmed from the work of the Climate Change and Me 1.0 (CC + Me 1.0) project. In that project, speculative fiction was introduced as a methodology that can enable young people to imagine possible worlds in response to the challenges of a changing climate and, more precisely, the changing material conditions of living in the Anthropocene (Rousell et al., 2017).

In the PhD study, speculative drama emerged as a methodology to creatively engage with young people in Australia and Sri Lanka in exploring alternative climate futures. This research study followed a participatory child-framed methodology (Barratt Hacking et al., 2013) which provided the co-researchers the platform to ecologically float in the space and to act as co-researchers and co-playwrights. The co-researchers participated in a series of online workshops involving the development and performance of their characters, scenes, environments, and plot lines within a collectively generated script (Figure 3.6). This process led to participants activating their climate-responsive initiatives in their local communities (Wijesinghe, 2022).

Speculative drama is an arts-based methodology developed during Wijesinghe's (2022) PhD research study. It is based on a posthuman theoretical framework and a participatory methodology that enables discussion through scriptwriting and performance (Wijesinghe, 2022). "Spaceulation" is a key concept that emerged during the speculative drama workshops and seeks to evoke what it means to think/with/in/through time and space (Wijesinghe, 2022, p. 211). The digital development of a script throughout the project suggested that speculating for the future involves more than just imagining and envisioning the future. What differentiates spaceulation from imagining and envisioning is the component of time that plays into the scripts written by the co-researchers. Thus, the final play script readings suggested the importance

FIGURE 3.6 The Co-researchers Reading the Scripts on Zoom Using the Digital Platform as a Medium of Expression to Perform Their Climate Change Thinking through Multiple Dimensions

and necessity of thinking through multiple dimensions of time and space into the future.

As a concept constructed through speculative philosophical thinking, spaceulation was generated as a creative term through this process. The data events that were identified in the "spaceulation" proposition depicted that the power of speculation bridges the gap between the present and the future and becomes a poly-temporal learning tool. Learning is embodied in the process of speculation and speculation engenders new forms of temporal experience (Wijesinghe, 2022, p. 233).

The concept of "spaceulation" holds interesting implications for understanding young people's digital media practices for climate justice because it emphasizes how multiple pasts and futures are continuously being reconfigured through the passing present as a contraction of differentiated "presencings." Digital media enable young people to co-mingle and interact with time and space by re-turning and returning to different pasts and futures, emphasizing the spatiotemporal multiplicity and plasticity of a general ecology. Social media technologies can accelerate this process in both generative and reductive ways. On the one hand, they provide new platforms for speculative temporalities and future-making practices. On the other hand, they establish new apparatuses of capture that "steal" their users' time, space, attention, data, and experience (de Freitas et al., 2020). Becoming aware of how digital media generate and capture multiple temporal and spatial dynamics is therefore key to empowering young people to use these tools in ethically responsive and socially transformative ways.

Speculative Methods for Youth Climate Activism through the Digital

In both examples discussed above, the network that forms between young people and digital technology is seen as a generator of possibilities that harness the importance of mobilizing social movements for climate justice. Both projects demonstrate how youth digital media practices can potentially shift the terms of value through which climate change is encountered and engaged, while establishing new forms of collective production that create rippling changes in the "general ecology" of Earth's planetary condition.

In this section, we discuss three methods that can be seen to operate as points of activation across both projects: *anonymous participation, character-building,* and *world-building.* The aim is to show how these methods involve a speculative and pedagogical outlook to address climate justice. Ultimately, we suggest that these new methods of educational response to climate change pave the way for new pedagogical tools and models to be developed. The speculative pedagogical approaches that each project took were derived through multiplicities created as reverberating effects from one project to another. We seek to continue and amplify this reverberating effect through this chapter in our ensuing discussion.

In both projects described above, strategies of *anonymous participation* emerged as key entry points for young people to engage with climate change through digital media. Young people opted to take part in the app design and script writing process by avatars and character names that they speculated during the various activities. They considered these speculated names to be more powerful than their own and even expressed that "future names" enabled them to speculate as a character rather than as themselves. Especially when the young co-researchers were collectively imagining the future world and the *app world*, they anonymized themselves to express their thoughts powerfully. This is consistent with increasingly widespread modes of anonymous participation in online chat forums and gaming platforms, which offer young people opportunities to contribute to digital spaces and events without disclosing their names and identities. This kind of anonymized participation has also become a crucial element of youth climate protest organizations, along with more radical climate protest initiatives which involve the disruption of urban transport networks and other illegal activities (Rousell & Chan, 2022). In these cases, youth activists often use encrypted messaging platforms such as Signal to mask their identities and evade capture by governmental agencies.

In the projects discussed above, anonymization enabled young people to express their thoughts as if they were not their own. The young co-researchers had the flexibility of anonymously moving from one point to another during the workshops, which enabled them to continuously rethink what the app, or what the future might look like, from different perspectives. As Wijesinghe (2022) writes, during the process of speculative drama, the co-researchers speculated about a time-space platform that was multidimensional (non-linear and in the formation of a general ecology), flexible, and inhabited by anonymous personas.

Similarly, in the construction of the Climate Action Adventure! App, spaces were explicitly constructed that enabled young people to post messages and organize climate activist initiatives anonymously through a masked Twitter account that couldn't be traced back to individuals (Figure 3.7). This area of the App also links directly to a Discord server set up specifically for global youth discussions and engagement with climate activism and education. Combining elements of "hacktivism" with speculative approaches to climate justice, these areas of the App connect young people's anonymous climate imaginaries with direct lines of climate communication and activism in the socio-political domain.

The second method that we identify across the two projects is *character-building*. In the CC + Me 2.0 project, young people's need to act anonymously

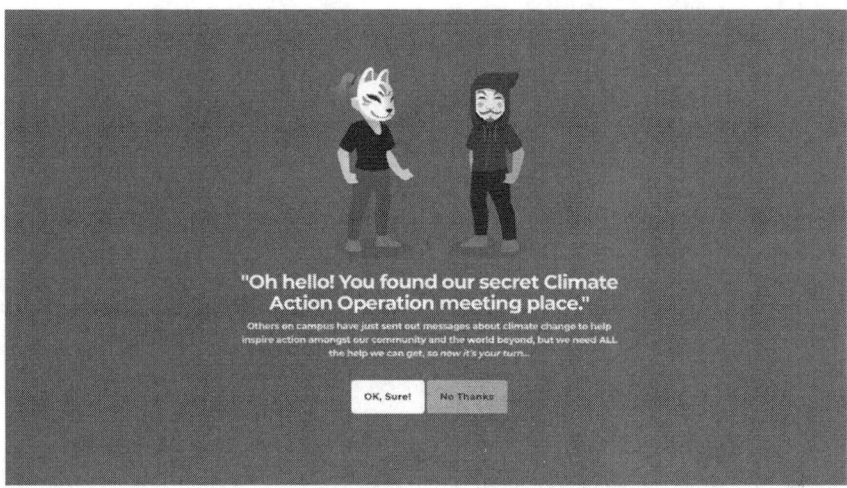

FIGURE 3.7   The Secret "Hideout" Uncovered below the Virtual Campus in Climate Action Adventure!, Which Invites Young People to Post-anonymous Messages about Climate Change into the Twittersphere

enabled them to create characters that they saw, envisioned, and speculated when looking at the future world. Thus, it was apparent that many characters were created in response to what they were imagining. In CC + Me 1.0, the co-researchers built speculative characters and some of these same characters were included by the CC + Me 2.0 co-researchers. The original characters were designed by 12-year-old co-researcher Jasmyne Foster in the original CC and Me project and were part of a collectively generated digital science fiction story in which young people develop mutant abilities that help them adapt to the disastrous impacts of climate change that are yet to come. These characters include Aoi, a young art student with a passion for climate science and environmental activism, and Elbereth, a young environmentalist with a passion for magic and the natural world. After a near-death experience, Aoi realizes that anything he draws in his notebook can come to life and help fight against climate injustice. After helping bring Aoi back to life, Elbereth discovers that she can draw on elemental energies to heal people and other creatures affected by climate change. Both characters became key elements in the Climate Action Adventure! App, which invites users to join Aoi and Elbereth on an adventure into climate justice worldwide.

Similarly, in Wijesinghe's (2022) speculative drama PhD research study, collective processes of imagining led to the creation of a play script by the co-researchers, which included several characters that they imagined through their speculations. A robotic professor who visits with his students from Mars to Earth in 2100 to study aspects of the mass-destructed Earth; a plastic bottle that is yet alive in 2100 and which was taken to Mars by a nature detective to examine in a Science Lab; and a half-human/half-robot who was saved by the professor all appear in the drama that speaks to what really happened to Earth. These characters emerge through the speculative stories the co-researchers narrated during the various speculative pedagogical activities in digital spaces. In some ways, it was as if young people were searching for stories in the ecological space of the digital, which also seemed as if it was their own world. In other words, the co-researchers were searching for their own characters in *this* world and also building their own new characters and worlds within the general ecology of the digital.

This takes us to the third method of *world-building* which embodies the "thousand ecologies" of young people's thoughts and ideas with concepts of fabulation and the minor politics of "a people and an Earth to come" (Deleuze & Guattari, 1994, p. 99). As a way of experimenting creatively with the emergent structure of alternative identities and possibility spaces, speculative pedagogies of world-building can generate social experiments that do not seek to transcend or suspend the political. Rather, they shift the political register into micropolitics and minoritarian resistance (Rousell et al., 2021). The digital world built in the CC + Me 2.0 project took the form of a speculative school campus. Once characters have been selected, the user is taken to a digital campus with its own cinema, library, art gallery, data dashboard, and live hashtag trackers for interacting with the latest climate science and activism campaigns. The cinema offers a selection of climate science, education, and activism films streamed from YouTube and watchable in a shared digital environment. The library contains a searchable database compiled from climate science data, modeling, narratives, and concerns from all over the world.

The potential value of building a virtual school became apparent through the features that the co-researchers built into this world during the design process. The space shuttle's dashboard connects data generated by in-App player activity with live updates of climate science data and live social media updates crowd-sourced from around the world. The "Top Tag Radar Report" continuously updates the most popular global hashtags connected to climate change activism. It allows young people to link to an aggregated collection of top-rated posts across multiple social media platforms. Other dials and readings on the dashboard are linked to current climate science data, adaptation and mitigation strategies, and modeling of future scenarios and tipping points. Launching missions from the dashboard is also available to explore how climate change affects Earth's environments.

Players can choose between four different "scapes" currently being impacted by climate change worldwide: Desertscapes, Forestscapes, Arcticscapes, and Cityscapes. As they explore each of these environments, players encounter "climate concerns" that have been crowd-sourced from children, young people, and adults through extended social media networks. Each climate concern focuses on a particular species or issues within a larger ecosystem, how this is impacted by climate change, and what can be done to address this concern. The "climate concerns" generation within the App is an ongoing project, leaving the world-building process open as we continue to field crowd-sourced submissions and responses from diverse global locations.

And the Stories Continue...

As we continue to analyze and explore the emergent insights and resonances between these two projects, we find new ways of connecting young people's speculative media practices with concrete levers of change in the physical world. In both cases, we find that co-researchers are effectively working to speculatively redesign their own education into a place where they can anonymously contribute to scientific, technological, and creative advances without being standardized, evaluated, and reduced to individualized capitalist subjects. In the design of Climate Action Adventure!, we see this playing out through Aoi and Elbereth's "secret hideout," which is hidden in the basement of an unmarked building within the App's virtual campus. We developed this space in response to the young people's desire to intervene in climate change discussions through various hidden identities. The space enables players to generate their own hashtags and post anonymously on Twitter using masked versions of Aoi and Elbereth as their avatars.

In the Speculative Drama project, where co-researchers collectively contributed to the ready-to-perform script, it was apparent that masking their identities enabled them to experiment creatively with the emergent structure of alternative identities and speculative possibilities for agency. This ability to think of new concepts and speculate about the future provided the young co-researchers the independence, autonomy, and confidence to become the designers of a future world. In addition, the speculative philosophical drama script enabled the co-researchers to speak collectively through characters other than themselves. For example, to voice the plastic bottle yet alive in 2100, the young co-researchers said that the plastic bottle had its own language and that they were translating it so that the adults would be able to understand it. The following excerpt from the drama script shares the thoughts of a plastic bottle that visits the Earth from Mars in 2100, along with the professor, to observe Earth after the mass extinction.

> I remember, a friend of mine back then, he was painted on by a human and sold to someone at a market fair. Then the person who purchased my friend fed him for a while with water, clean water, and then took him on a spinning ride by throwing him into the ocean (laughs) he landed next to me.
>
> (*Wijesinghe, 2022, p. 166*)

From the perspective of general ecology, the thoughts that co-researchers attribute to nonliving things are not simply a personified projection or anthropomorphic characterization. Rather, they can be understood as a speculative translation of how young people are thinking-feeling the current conditions of the world in relation to all other things that exist. In the co-researcher's words, the only way to understand the world through others' perspectives is by "listening carefully to the things, understanding it, and then translating the feelings so that every human being understands the story behind an Earth destructive plastic bottle, or a tree that was just cut due to industrialisation" (Wijesinghe, 2022, p. 135). Digital media provides a significant platform for collectively developing and sharing speculative insights and ways of coming to know the world in a changing climate.

Considering these speculative digital projects as sites of generating knowledge differently, it becomes clear that education requires an urgent change to become worthy of the general ecological conditions of the Anthropocene. Could digital media provide the platform for new pedagogies and curricula to be actively co-developed together with young people? Could youth inputs enable the education system to provide solutions to issues like climate change by valuing young people as active contributors to a future that's failed to be realized by adults? Is it time that every educational conference, every educational board, and department, and every education policy and budgeting initiative included children and young people's voices and experiences to collectively reform education policies and practices?

We believe that it is time for these questions to be considered as global education systems face the "certainty of ecocide" and the implications of severe climatological disruption (Finn & Phillips, 2023). Such questions gesture toward a radical overhaul of an education system based on archaic models of standardization and hyper-individualization, a system unfit to address the collective challenge of responding to climate change at both global and local scales. General ecology and related philosophies of relationality and interdependence provide agile theoretical models for re-imagining education at the systemic level. Ultimately, young people's lived experimentation and articulation of such theories can establish new terms of value for living and learning under conditions of the climate crisis.

References

Barratt Hacking, E., Cutter-Mackenzie, A., & Barrratt, R. (2013). Children as active researchers: The potential of environmental education research involving children. In R. Stevenson, A. Wals, M. Brody, & J. Dillon (Eds.), *The handbook of research on environmental education* (pp. 438–458). American Educational Research Association.

Boulianne, S., Lalancette, M., & Ilkiw, D. (2020). "School strike 4 climate": Social media and the international youth protest on climate change. *Media and Communication*, *8*(2), 208–218. doi:10.17645/mac.v8i2.2768

Brunner, C., & Fritsch, J. (2011). Interactive environments as fields of transduction. *The Fibreculture Journal*, (19). https://eighteen.fibreculturejournal.org/2011/10/09/fcj-124-interactive-environments-as-fields-of-transduction/

Caron, C., Raby, R., Mitchell, C., Thewissen-LeBlanc, S., & Prioletta, J. (2017). From concept to data: Sleuthing social change-oriented youth voices on YouTube. *Journal of Youth Studies*, *20*(1), 47–62. doi:10.1080/13676261.2016.1184242

Cull, L. (2013). Philosophy as drama: Deleuze and dramatisation in the context of performance philosophy. *Modern Drama*, *56*(4), 498–520. doi:10.3138/md.S87

Cutter-Mackenzie, A., & Rousell, D. (2018). Education for what? Shaping the emerging field of climate change education with children and young people as co-researchers. *Children's Geographies*, *17*(1), 90–104. doi:10.1080/14733285.2018.1467556

de Freitas, E. (2020). Why trust science in a trickster world of absolute contingency? The speculative force of mathematical abstraction. *Critical Studies in Teaching and Learning*, *8*(1), 60–74. doi:10.14426/cristal.v8iSI.278

de Freitas, E., Rousell, D., & Jager, N. (2020). Relational architectures and wearable space: Smart schools and the politics of ubiquitous sensation. *Research in Education*, *107*(1), 10–32. doi:10.1177/0034523719883667

de Freitas, E., & Truman, S. E. (2021). New empiricisms in the Anthropocene: Thinking with speculative fiction about science and social inquiry. *Qualitative Inquiry*, *27*(5), 522–533. doi:10.1177/2F1077800420943643

Deleuze, G. (1967). The method of dramatisation. *Bulletin de la Société française de Philosophie*, *LXII*, 89–118.

Deleuze, G., & Guattari, F. (1994). *What is philosophy?* Columbia University Press.

Finn, R., & Phillips, L. G. (2023). On the certainty of entanglements with ecocide: Pragmatic action for responsive pedagogy inspired by ecological psychology and permaculture. *Educational Review*, *75*(1), 115–133.

Fritsch, J. (2011). Affective experience in interactive environments. *The Fibreculture Journal*, (19). https://nineteen.fibreculturejournal.org/fcj-137-affective-experience-in-interactiveenvironments/

Fullagar, S., Rich, E., & Francombe-Webb, J. (2017). New kinds of (ab)normal?: Public pedagogies, affect, and youth mental health in the digital age. *Social Sciences*, *6*(3), 99. doi:10.3390/socsci6030099

Gabrys, J. (2020). Smart forests and data practices: From the internet of trees to planetary governance. *Big Data & Society*, *7*(1). doi:10.1177/2053951720904871

Hansen, M. B. (2015). *Feed-forward: On the future of twenty-first-century media*. University of Chicago Press.

Hörl, E. (2017). *General ecology: The new ecological paradigm*. Bloomsbury Publishing. ISBN: ePDF: 9781350014688

Maher, T. V., & Earl, J. (2019). Barrier or booster? Digital media, social networks, and youth micromobilization. *Sociological Perspectives*, *62*(6), 865–883. doi:10.1177/0731121419867697

Manning, E. (2012). *Relationscapes: Movement, art, philosophy*. MIT Press.

Manning, E. (2013). *Always more than one: Individuation's dance*. Duke University Press.

Manning, E. (2023). *Out of the clear*. Minor Compositions.

Massumi, B. (2002). *Parables for the virtual: Movement, affect, sensation*. Duke University Press.

Massumi, B. (2015). *The politics of affect*. Polity Press.

Mayes, E., & Holdsworth, R. (2020). Learning from contemporary student activism: Towards a curriculum of fervent concern and critical hope. *Curriculum Perspectives*, *40*(1), 99–103.

McGillivray, D., McPherson, G., Jones, J., & McCandlish, A. (2016). Young people, digital media making and critical digital citizenship. *Leisure Studies*, *35*(6), 724–738. doi:10.1080/02614367.2015.1062041

McLean, J. E., & Fuller, S. (2016). Action with (out) activism: Understanding digital climate change action. *International Journal of Sociology and Social Policy*, *36*(9/10), 578–595. doi:10.1108/IJSSP-12-2015-0136

Protevi, J. (2013). *Life, war, earth: Deleuze and the sciences*. University of Minnesota Press.

Ringrose, J. (2011). Beyond discourse? Using Deleuze and Guattari's schizoanalysis to explore affective assemblages, heterosexually striated space, and lines of flight online and at school. *Educational Philosophy and Theory*, *43*(6), 598–618. doi:10.1111/j.1469-5812.2009.00601.x

Rousell, D. (2019). Inhuman forms of life: On art as a problem for post-qualitative research. *International Journal of Qualitative Studies in Education*, *32*(7), 887–908. doi:10.1080/09518398.2019.1609123

Rousell, D., & Chan, K. K. L. (2022). Activist ecologies of study in the learning city: Deformalisations of educational life. *Discourse: Studies in the Cultural Politics of Education*, *43*(5), 702–722.

Rousell, D., Cutter-Mackenzie, A., & Foster, J. (2017). Children of an earth to come: Speculative fiction, geophilosophy and climate change education research. *Educational Studies*, *53*(6), 654–669. doi:10.1080/00131946.2017.1369086

Rousell, D., & Cutter-Mackenzie-Knowles, A. (2023). *Posthuman research playspaces: Climate child imaginaries*. Routledge.

Rousell, D., Wijesinghe, T., Cutter-Mackenzie-Knowles, A., & Osborn, M. (2023). Digital media, political affect, and a youth to come: Rethinking climate change education through Deleuzian dramatisation. *Educational Review*, *75*(1), 33–53.

Sholtz, J. (2016). Dramatisation as life practice: Counteractualisation, event and death. *Deleuze Studies*, *10*(1), 50–69. doi:10.3366/dls.2016.0211

Taylor, E., & Rooney, T. (2016). Digital playgrounds: Growing up in the surveillance age. In E. Taylor & T. Rooney (Eds.), *Surveillance futures: Social and ethical implications of new technologies for children and young people* (pp. 1–16). Routledge.

Verlie, B. (2019). Bearing worlds: Learning to live-with climate change. *Environmental Education Research*, *25*(5), 751–766. doi:10.1080/13504622.2019.1637823

Wijesinghe, T. (2022). An exploration of how Speculative Drama can be engaged to understand how Young People speculate about their future worldviews on Climate Change [Unpublished doctoral thesis]. Southern Cross University.

4

INTEGRATING COMMUNITY-BASED PARTICIPATORY RESEARCH APPROACHES WITH CLIMATE JUSTICE DIGITAL MEDIA PROJECTS

Emily Polk

Introduction

My student approached me early in the quarter with an idea for a research project she was very excited about. As a geology major focused on heavy metal pollution variability, she wanted to test for soil contamination in underserved communities, most of whom were Latinx, Black, and indigenous near our university. She was eager to learn the community-engaged research methodologies and writing practices that we explore in my Intro to Environmental Justice course that I co-teach with research scientist Dr. Sibyl Diver. Her intentions were well-meaning and grounded in the histories of environmental injustice we center in our class.

She knew, for example, that the same industries that drive climate change cause disproportionate harm to underserved and marginalized communities. Toxic waste dumping, disposal wells from fracking, and hazardous landfills have been placed disproportionately in communities with high proportions of people of color and residents living in poverty (Mohai & Saha, 2015). Moreover, polluters and high levels of pollutants, including automobile fumes, smog, soot, oil smoke, ash, and construction dust, all of which are linked to serious health problems, are disproportionately located in communities of color (Mikati et al., 2018). In fact, a 2017 report by the National Association for the Advancement of Colored People and the Clean Air Task Force found that African-Americans are exposed to 38% more air pollution than white people and are 75% more likely to live near toxic pollution than the rest of the American population (Fleischman & Franklin, 2017).

DOI: 10.4324/9781003335276-6

It's not just toxic waste dumping. Research also shows that lead poisoning, water contamination, and displacement disproportionately impact communities of color and low-income communities. More recently, the Harvard T.H. Chan School of Public Health published data linking air pollution to higher COVID-19 death rates. Looking at 3,000 counties, they found that someone who lives for decades in a county with high levels of fine particulate pollution is 8% more likely to die from COVID-19 than someone who lives in a region that has just one unit ($1 \mu g/m^3$) less of such pollution (Wu et al., 2020).

Even though my student did not have a relationship with any community organizations working on health and climate justice issues, she assumed that her project would be embraced because her research would be "helping" them. "But what happens if you find contaminants in the soil that exceed EPA standards?" I asked her. "What if your research ends up forcing families to uproot and relocate? Do you know the housing policies and regulations for these communities?" I asked her if her research design included how she would share resources for remediation and tips for getting support should high levels of contamination be found. Would she cultivate relationships with any of the people whose soil she was testing? How would she do that? Would her project offer the next steps for ensuring the community's health, or would she only be sharing her scientific findings?

I asked these questions because I want my students to understand how the implications of their findings are connected to the decisions they make about communicating them. I wanted them to see how science is embedded in and connected to a set of social, cultural, and political systems, and how, when, and where scientists share those findings have tremendous consequences—positive and negative—especially when the project centers on environmental and climate justice issues. Sometimes, the most well-intentioned research projects can be the most extractive and produce the most harm.

Teaching Environmental Justice Communication

As a teacher at an elite STEM university, most of my students, who range from freshman to graduate students, will begin careers in the sciences when they graduate. They take my environmental writing classes to have the opportunity to engage collectively and critically in the challenges and possibilities of doing community-engaged environmental justice research. They also want to develop the skills to communicate findings via a range of media projects that consider the intended audience, purpose, and impact. These two goals are grounded in teaching students how to communicate about the climate crisis through a justice lens that aligns with the communities made to bear the brunt of environmental harm.

Students in my classes have:

- Built a digital storymap that integrated US Census data to examine the spatial distribution and demographic context of the 2016 Portland Public School (PPS) lead readings.
- Partnered with the research team at the nonprofit Communities for a Better Environment to create infographics and PowerPoint presentations to aid in strategy development regarding their campaigns to shut down a Chevron refinery.
- Developed a digital children's book featuring a youth climate justice activist from Los Angeles and a comic for Indian youth that focused on the challenges and victories of the Jal Sahelis, a community of rural women navigating severe drought.
- Partnered with Kīpuka Kuleana, a nonprofit on the North Shore of Kaua'i working to resist land dispossession and displacement, to launch a data visualization project that mapped recent land grabs by wealthy tech CEOs.
- Partnered with community organizations in the central valley of California to create an interactive and accessible web mapping tool of oil wells in California, among many other digital projects.

While all the projects had different goals, contexts, and audiences, each was grounded in a similar ethos. Students approached their research *not* with the intention of "saving their subjects" from climate change but rather just the opposite—they were challenged with doing the hard work of saving *themselves* from repeating the potential cycles of harm caused by extractive research by using the writing and research process to practice environmental justice (hereafter EJ) principles (The Principles of Environmental Justice, 1991; *ejnet. org/ej*) and ethical community-engaged approaches (Haas Center for Public Service, Stanford University; *haas.stanford.edu*): humility, mutual respect, and reflexivity. This included understanding how their positionality impacted their research questions and approaches, and thoughtful preparation—including learning about the social and cultural contexts that ultimately informed both the focus of their projects and their approach to building relationships and trust within communities.

Applying these principles to students' research projects also helped them think through the most effective digital mediums to communicate their work because they were listening and responding to the needs and desires of the communities in their projects. In this way, our class was as much about the process of teaching how to communicate EJ research effectively, as it was about how to be an ethical and empathetic human in a world on fire.

Because our EJ class is not a technical class—e.g., we do not teach students the latest video technologies or podcasting techniques—this chapter has been organized based on the structure of our course, with references to the diverse student digital media projects throughout.

Step 1: Starting Their Research Projects with the Histories They Need to Know

My students are fueled by the urgency of the climate crisis and seek to use their projects to make meaningful interventions. However, they must understand that they are entering a research conversation that has historically been marked by extraction and exploitation of marginalized people in the form of "helicopter research."

A recent article in *Nature* describes helicopter research as occurring when "researchers from high-income settings, or who are otherwise privileged, conduct studies in lower-income settings or with groups who are historically marginalized, with little or no involvement from those communities or local researchers in the conceptualization, design, conduct or publication of the research" (Adame, 2021). Fernanda Adame, a Research Fellow working at the Australian Rivers Institute at Griffith University, notes, for example, that famed British biologist Charles Darwin used his research in South America to write a book foundational to modern biological science, *On the Origin of Species*. However, Darwin rarely mentioned the people whose ancestors had lived in the Americas for 2,500 years, "except a few instances when Darwin refers to the Indigenous people of Tierra del Fuego as 'savages' or 'barbarians'." She makes the important point that this kind of helicopter research was not rare and continues to this day.

The practice of scientists from wealthy nations visiting lower-income countries, collecting samples, publishing the results with little or no involvement from local scientists, and providing no benefit to the local community…was not exclusive to Darwin: during the nineteenth century, German explorer Alexander von Humboldt, who visited Venezuela, Colombia, Cuba, and Mexico, and British naturalist Alfred Russel Wallace, who collected biological specimens in southeast Asia, did the same. And even now, helicopter research is still common in science (Adame, 2021, np).

A recent editorial in the prestigious journal *Nature* highlights insightful examples, including one analysis of a sample of studies conducted in Africa on infectious diseases that found that less than half had an African first or last author. Another report showed that two-thirds of high-impact geoscience articles on Africa had no African authors (Mbaye et al., 2019). Another study found that in development research, authors from the global north wrote nearly three-quarters of papers in the world's top 20 development journals between 1990 and 2019, even when the focus was overwhelmingly on challenges facing the global south (North et al., 2020).

An essential part of teaching students how to intervene effectively in the climate crisis is introducing them to how these historically colonial approaches are still a part of how we approach problem-solving today. Ultimately, sharing this historical context helps students understand the importance of ensuring that these community voices are not erased. They could then give credit to these voices in their knowledge production and disrupt erasure cycles by centering their community expertise and knowledge in their climate justice research projects.

Community-based participatory research (CBPR) is one meaningful intervention that is particularly useful for students in writing and rhetoric classes focused on climate change. CBPR is a partnership approach to research that equitably engages with and builds upon the strengths and resources of a community in all aspects of the research process. It also ensures that the community shares control of and benefits from the research. The concept of community as an aspect of collective and individual identity is central to CBPR. Community is characterized by identification with other members, common symbol systems, shared values and norms, mutual influence, common interests, and joint commitment to meeting shared needs. Communities of identity may be geographical or consist of members of a geographically dispersed group with a sense of common identity—shared ethnicity, sexual orientation, etc. (Israel et al., 2008).

My students certainly cannot be expected to develop and implement a CBPR project in a short ten-week writing class. However, we introduce the principles of community-based "participatory research as an orientation to research that aims to help them critically engage with who has the right to speak, to analyze and to act" (Hall, 1992, p. 22). Their critical engagement informs their rhetorical choices at their research design, the sources they include as "experts," and the digital medium they choose to communicate their research—regardless of whether or not they have the capacity to engage directly with a community.

Engaging with the principles of CBPR and the histories from which they emerged is central to any long-term work on climate justice. I hope my students will take these approaches outside our classroom and into their careers. Any effort to develop equitable long-term solutions to climate change must also work to change the inequitable social and economic systems that have ensured that communities of color and lower-income communities are made to bear the brunt of climate impacts and environmental harms (Polk et al., 2023).

Such systems thinking entails analyzing whose voices have the most power to create and perpetuate the systems. Whose voices are missing from the table where solutions are developed? As a positive example of a successful effort to change extractive systems, I share the work of the Climate Justice Alliance (CJA; *climatejusticealliance.org/about*). CJA was formed in 2013 to unite frontline communities and organizations and now has more than 80 urban and rural frontline communities, organizations, and support networks in the climate justice movement.

Their trans-local organizing that places race, class, and gender at the center of their work is building a Just Transition away from extractive systems of production, consumption, and political oppression toward resilient, regenerative, and equitable economies. They have won significant victories against polluting and extractive industries, preventing new carbon emissions and building local alternatives that center on traditional ecological and cultural knowledge.

Step 2: Project Brainstorming Begins with Connecting EJ Projects with Interests, Other Coursework, and Extracurricular Activities

Managing expectations is crucial for any class focused on developing and completing a research-based project. It is especially so in a ten-week class filled with overachieving students who see the intersections of our multiple global crises—climate, health, economy—and are eager to make a meaningful contribution to ensuring justice and equity in the movements to address climate change. We make sure students understand that our class can be the beginning of a much longer project. They are not required to rush and complete a final assignment but rather to develop manageable goals relative to our time together and work to do as complete a job as possible within the constraints of our limited time frame. We do this by scaffolding the big project into smaller assignments, in-class activities, and discussions.

Students have used our class to lay the foundation for longer-term projects and as an opportunity to translate EJ work started in other classes into digital media projects. Many students have started projects in their other classes that feature a component of EJ. However, they are not necessarily focused on distributing and disseminating their work in those classes—something they can do in our class. We want them to understand EJ as an approach and an orientation to problem-solving around the climate crisis, not only as a singular course distinct from their other academic experiences. Thus, we encourage students to use our writing exercises, readings, in-class discussions, peer reviews, and lectures on EJ theories, approaches, and frameworks to connect with their other coursework, extracurricular activities, and previous internships and jobs.

For example, my student Evan was the president of Students for Sustainable Stanford (SSS) when he was a student in our class. SSS is the oldest and most prominent student-run organization focused on long-lasting sustainable practices on and off Stanford's campus. SSS works on climate change, EJ, water, waste, and more, with a mission to educate the broader Stanford community and connect students to sustainability projects. Evan used our class time to create a digital presentation (which looks like a slide deck) with accompanying audio recordings to help what he called "mainstream environmental college groups" integrate EJ principles and practices into their organizations. In addition, he used his research to help inform SSS's initiatives and share the project with different students and environmental college groups around the country.

My student Andrea decided to use our class to create a "Just Preparedness Digital Toolkit" to help support the frontline community organizers in her native Puerto Rico during the pandemic. She wanted to facilitate community building digitally when people could not connect physically. Her toolkit includes need-assessment surveys, maps of community resources, and a community skills spreadsheet, among other resources. One of her main objectives for digital community-building recommendations was to ensure the accessibility of the tools. She created a hybrid between a curriculum and a toolkit hosted on the Google Drive platform. She designed this toolkit for the community organizers she was already working with to help them facilitate connectivity and support.

Although the toolkit was on a (free) digital platform, she designed it to be accessible regardless of technical expertise. She was already aware of the telecommunication constraints in Puerto Rico. However, she knew that most households, especially in the regions she was working in, have some access to Wi-Fi and service, particularly cell phones, and Google Drive has mobile apps for all its services. Based on her experiences and local community connections, she knew that the platform would be easy to share and collaborate with others. It also provided access to maps that display desired data layers and locations, customizable surveys, and data sheets that would help build the infrastructure to help support well-functioning communities.

Step 3: Positionality and Reflexivity

Evan and Andrea were embedded in communities centered on their projects, the former in his academic community and the latter in her hometown community. As a result, they understood how their positionalities informed their research purpose, focus, and design. Understanding positionality is a core concept for community-engaged work, and it is a central feature of our class. The concept of positionality is based on a long history of post-Marxist feminist geographers (Haraway, 1998; Harding, 2004; Rose, 1997). It describes how an individual's perspective is shaped by their social position, including class, gender, sexuality, racial identity, and other determinants of social privilege (Pulido & Peña, 1998).

As part of their research orientation, we ask students to participate in acknowledging and deconstructing the dominant narratives and personal privileges embodied in their race, class, gender, and other identities. Doing so helps them see how their multiple positionalities have shaped their understanding of the world. It also leads to generating the questions they want to ask; the sources they have deemed to be experts; and recognizing how people experience and respond to environmental problems differently based on their social position.

Many of my students want to engage with power dynamics as they have contributed to and sustained environmental injustice. Our positionality work in the class asks students to be accountable for their own power. We do this in two ways with our students. (1) We ask them to "situate themselves" as part of their first assignment: a brief research proposal. More specifically, we ask them to briefly state how their personal, intellectual, academic, or work background and/or experiences have led them to this particular research project and why it is important to them. (2) We do a small group in-class activity that asks students to discuss with each other their positionality as a researcher.

They also discuss how their various identities intersect with the questions they have asked; how they might influence the way sources are selected and interpreted; and what affordances, vantage points, and/or blinders they might have. To be clear, a positionality statement was not necessarily part of the final project. However, the exercises informed and complicated their thinking and research approaches, and, for some, the capacity to communicate with the communities involved in their projects.

For example, my student Natalie created the digital ArcGis Storymap platform to explore the role of music in climate justice movements. Storymaps allow for the creation of immersive stories by combining text, interactive maps, and other multimedia content. Natalie used the platform because it allowed her to embed music videos so that readers could listen alongside reading her piece. The platform was also conducive to a more informal format that allowed her to communicate her positionality in the piece. The opening page of Natalie's storymap reads:

> Music has been a constant in my life for as long as I can remember. Whether it be dancing to Rihanna in my living room or rocking the car back and forth to the beat of Bob Marley; the melodies and rhythms of Caribbean music are a central component of my connection to my family. My mom has told me time and time again that our shared calypso Spotify playlist is one of the most tangible ways she is able to feel connected to her island home, Barbados, despite living thousands of miles away landlocked in the middle of the United States. These transcendent powers of music are not only integral in strengthening the connections between diasporic community members and their homelands, but also leads to impactful results when musical mediums are leveraged for activism.

After situating her personal connection to her topic, she contextualized its significance in the context of climate change, writing,

> This activism is essential as the Caribbean is currently in the midst of one of the most urgent and life-threatening issues—the climate crisis... This is nothing new, however a 2018 report by the International Panel on Climate Change (IPCC) *ipcc.ch/sr15* published overwhelming evidence that an increase of over 1.5°C in global temperatures will result in irreparable damage to many of these frontline Caribbean nations...For the people of the Caribbean, climate change is not a future threat but rather a very real matter of life or death.

Natalie notes that as a student majoring in Earth Systems, this is a topic that not only aligns with her academic interests but is also deeply personal. "The people living on these Caribbean islands are not just statistics, but rather my family members and friends."

Similar to Andrea, Natalie's positionality reflected her connections to her native community. However, not all students develop projects connected to their local communities. My student Heather researched and wrote a policy brief to help inform policy decisions around lead paint poisoning in Oakland, California. She did not grow up in Oakland but had a strong interest in supporting policy reform there. In her proposal, she used her positionality statement to practice humility and to think through where her potential blinders might be, an exercise that helped her communicate more effectively when reaching out to community leaders. She wrote in her proposal:

> I grew up in Palo Alto, California, and was able to enjoy the many benefits of living in the Bay Area. However, as I grew older...I began to become aware of local social injustices. This summer, working in environmental engineering consulting in Oakland, I was able to expand my perspective even further. However, as an educated, white, middle-class worker, I was disconnected from a large part of the Oakland community. The environmental issues within the Oakland area were left out of the main narrative I saw at work... I understand that given my privilege and my place as a white, affluent, outsider I will never fully be able to understand the intricacies of their community struggles. I know that I will naturally draw my attention to traditional forms of scientific evidence and institutional power structures. I will likely miss locally-valued issues, and I hope to engage humbly and extend my understanding through direct communication with community members and people who come from a different positionality.

Step 4: Making Connections with Community Voices

Students have a range of relationships with the communities featured in their projects. Some relationships are closer and require more time commitment; some involve lighter forms of involvement. The range typically depends on the type of preexisting relationship (if there is one) and a student's capacity to build a long-term relationship. We encourage students who are reaching out to a community group or leader for the first time to

1 First, identify the need/community need and find out if they have a mutual friend, teacher, or mentor who can introduce them. Cold calls are often less effective than working with a mutual connection, where there is already a level of trust.
2 Write a short email that (a) indicates familiarity with and respect for the groups' work/efforts, and how they learned about them and (b) *briefly* introduces who they are and what they are doing, including their identity as a student and any skills/background that is relevant, (c) contains a clear and doable action item (e.g., do they have a staff member who could speak to them about this issue; is there time for a short conversation in the next week or two, could they recommend materials that describe more about their organization's approach on these issues?), which respectfully acknowledges time limitations and thanks them in advance for their time.

My student Cindy did not have a previous connection with the famous youth activist Nalelli Cobo, who won the prestigious Goldman Environmental Prize in 2022 (*goldmanprize.org/recipient/nalleli-cobo*). Cobo fought against and eventually shut down oil drilling in her LA neighborhood. However, when Cindy read about her story, she knew that she wanted to create a project to make the activist's story more accessible to a younger audience. Cindy wrote in her reflection:

> I decided that the story would target readers between the ages of five and eight. Readers should walk away from the story with an elementary understanding of the history of oil drilling in Los Angeles, the impact Allenco's oil drilling had on University Park residents, and how Nalleli and her community responded to it. While the story is about how a community overcame environmental injustice, what it's really about is how ordinary youth can do extraordinary things.

When Cobo was just nine years old, she began suffering from heart palpitations, body spasms, and severe nosebleeds. She learned from a toxicologist about the environmental and health hazards of urban oil drilling being emitted from less than 50 feet away from her home and began to connect the dots. As an elementary school student, Cobo knocked on doors with her mother, attended city hall hearings, and fought for the termination of oil drilling in her community. Their advocacy got the attention of state officials, and by the time she began high school, Cobo had helped shut down Allenco's oil operations, becoming a local hero.

Cindy wanted to create a digital children's book about Cobo's story but wanted to ensure that she had the young activist's input and approval. After conducting the bulk of her research, she reached out to Cobo and interviewed her to learn more about the details of her story. Cindy wrote in her reflection:

> As I interviewed Nalleli, I focused on elements of the story that would help me craft a compelling and visual narrative. I asked her when she found the time to knock on doors with her mother. I asked her where she was when they found out the oil drilling had been ordered to shut down. Collectively, details such as these helped bring the story to life in a way the articles I read about her did not.

After completing most of the illustrations that would accompany the text of the children's book, Cindy shared the rough draft of the book with Cobo to ensure that she presented her story in a way that she felt was true to her experience. Cindy wrote:

> I wanted to make sure she was comfortable with the micro decisions I made throughout the book, from the way I condensed her story for a younger audience to the way I illustrated her and her family. Nalleli responded favorably and gave me constructive feedback on a part of the story that came across as factually incorrect, which I used to clarify the final text...I was pleased to hear that reading the draft was a touching experience for both her and her mother. I hoped it meant that I at least partially succeeded in centering their voices and experiences in a way that felt authentic and moving to them.

Cindy's effort to directly contact the subject of her research even though she didn't have any prior relationship is just one pathway students utilize. Typically, there are four pathways students take for making connections with communities. (1) They already have a connection with a community organization because they have worked with them in some capacity, either as an intern or volunteer or as part of another research project in a different class. For example, my student Claire came into our class while researching the oil and gas industry with a graduate student, who was able to facilitate her connections to two community-based EJ organizations in California. She used class time to develop a mapping tool with input from the organizations to create a data-driven web mapping tool that allows community members to assess their exposure to oil and gas wells (*vision-ca.org*).

Claire developed the tool in collaboration with activist Kobi Naseck, coalition coordinator for the Voices In Solidarity Against Oil in Neighborhoods (VISIÓN) Coalition (their mapping tool is published on their website *vision-ca.org*). She also worked with Dan Ress, an attorney with the Center on Race, Poverty, and the Environment (CRPE). They both helped her design the mapping tool and shared information community members would find useful on the final map. She wanted to use our class time to finish the design of the mapping tool with input from our EJ community and to have the opportunity to reflect on the links between quantitative research (her specialty), policy, and advocacy.

My student Sydney had a previous connection to a community org. She decided to create a presentation and storymap *t.ly/VSK3* to share with the Henry's Fork Foundation's (HFF) 30 staff members. She had spent the previous summer working as an Education and Interpretive Center Intern with HHF, a nonprofit based in eastern Idaho focused on protecting and restoring the fisheries and wildlife of the Henry's Fork and its watershed, and a member of the Henry's Fork Watershed Council (HFWC). Sydney was concerned that the voices and expertise of the Shoshone-Bannock people, who were indigenous to the area and the original stewards of the watershed, were excluded from representation in the HFWC and scientific literature discussing the collaborative management of the Henry's Fork.

Sydney noticed that the Shoshone-Bannock were only mentioned once on the HFF website, and its museum had only a small museum placard. She created a storymap as a multimedia exploration of the history of the tribes in the area and their role as caretakers of the river. The storymap format allowed Sydney to integrate videos made by the Shoshone Bannock to center their voices. Her connection to the HFF organization allowed her to facilitate a presentation that featured the storymap and included time for collaborative discussion and learning. She explained her choice to use the storymap format:

> I thought that utilizing a formal research paper would not properly honor the voices of the Shoshone-Bannock people from which my research centered or acknowledge the different forms of knowledge and content produced by the tribes...it also felt wrong to manufacture or alter Indigenous voices to fit within a western scientific context, and also alter Indigenous voices which would be problematic due to my positionality as a non-Indigenous person.

The second pathway students take in developing connections with communities is by connecting with the communities with which we (the instructors) have relationships and/or via the leaders we invite to campus through our annual EJ speaker series. In an effort to platform the knowledge and expertise of frontline communities in our class, we invite EJ leaders to guest teach on a range of issues, including climate justice, food justice, and queer ecologies, Afrofuturism, indigenous knowledge, toxic waste exposures, among other topics (Polk & Diver, 2020). For example, students working on projects related to food justice have interviewed speaker Haleh Zandi of Planting Justice to integrate her work supporting formerly incarcerated individuals as part of an intentional and diverse urban agriculture community.

Several students reached out to the organizers at Communities for a Better Environment (CBE) *cbecal.org* after their visionary leader Andres Soto shared in class the long history of environmental injustice endured by his community in Richmond. CBE has a long history of resisting racism and working to reduce pollution and build green, healthy, sustainable communities. For example, my student David followed up with CBE and worked with them to create a digital factsheet to persuade the Richmond Air District to require the Chevron oil refinery to reduce PM emissions by 160 tons per year. He also created a community-oriented factsheet to raise awareness and encourage community engagement on the issue.

My student Olivia also collaborated with CBE. She developed a PPT presentation for CBE that explored the Philadelphia Refinery Complex explosion and closure. When she took our class, CBE was in a research and strategy phase in their campaign to shut down the Chevron refinery. It was important to them to have research support regarding the clean-up and land remediation needed when the refinery closed, including assurance that Chevron would be held accountable for the clean-up. CBE knew that this struggle for accountability for land remediation of a refinery site was playing out in South Philly between community organizers, Philadelphia Energy Solutions and Sunoco (the former owners of the refinery site), and Hilco Redevelopment Partners (the current owners of the site).

CBE needed a researcher to help them use the South Philly case to inform CBE's strategy around what the accountability for clean-up might look like in Richmond. Olivia used the case study to look at organizing tactics and strategies used by organizers in Philadelphia. According to Olivia, CBE requested, "an annotated notes document with clickable links for everything [I] find, and a ~30 slide-long slide deck (visual and easy-to-follow for staff and community members) for everything." Olivia's presentation served a dual purpose: for staff, it helped to guide further research into legal strategies and organizing tactics relevant to hold Chevron accountable for clean-up, and for community members, the presentation situated the campaign in Richmond in relation to the campaign in Philadelphia in order to provide more context and understanding of the objectives of CBE's/Richmond Our Power's Just transition campaign.

My co-instructor Sibyl Diver helped our student Laura to connect with the Aborigen Forum, an informal alliance of expert activists and leaders and community organizations representing indigenous peoples and community leaders of the North, Siberia, and the Far East. Sibyl, a community-engaged scientist, focused on indigenous governance, had a longstanding relationship with them. Laura developed a research paper and PowerPoint presentation to support the forum's campaign against Nornickel, a mining and smelting company that is one of the largest producers of nickel and one of the biggest polluters of the Arctic. The company has a history of environmental harm, particularly in indigenous communities.

The third pathway for community engagement for our students is via their family members, friends, neighbors, and colleagues, many of whom are from communities that have been made marginalized. The opportunity to feature family members and friends as experts and knowledge producers and creators in their projects provides students with a sense of validation and agency. For example, my student Stephanie, a first-generation Afro-Costa Rican, created a podcast with the help of elders in her family that told her family's history in Limón, Costa Rica about EJ concerns in Costa Rica. She argued that knowing about transportation is important to understanding our current problems and as one lens "that gives us the language and tools to pave a 'better road' forward for our communities."

Finally, the fourth pathway is utilized by students who want to engage with community standpoints and interests but cannot engage directly with a community organization. For example, if they are not able reach out to a central contact the way Cindy did with her children's book, we encourage them to access and share community voices through numerous digital avenues. This includes but is not limited to listening to recorded public testimony; tracking social media posts; drawing on direct quotations in online publications: analyzing frontline community org website materials and recorded online talks (including our guest speaker).

Step 5: Choosing the Digital Media Format: Activities to Think Through Audience, Purpose, and Impact

Through these projects, students learn the historical context for EJ movements to engage individually and collectively with EJ principles and community-engaged approaches to research through their project proposals. In doing so, they are reflecting on the affordances and challenges of their positionality relative to their research projects. We then provide several activities to help them think through the best genre to communicate their project.

First, it is helpful if they understand what the word "genre" means. We introduce an activity adapted from my colleague Shay Brawn that helps students to understand how the formal features, stylistic conventions, and content choices associated with genres arise out of largely predictable (although perhaps evolving) constellations of audiences, exigences, and constraints. So, for instance, we can see the values and purposes shared by scientists embodied in the genre of the scientific research article in its formal features (abstract, introduction, methods, results, discussions), its characteristic style (concise, precise, impersonal, referential), and its contents (data in various forms, details about experimental and analytical methods).

We ask students to think deeply about how the features, styles, and contents relate to the kinds of purposes a genre serves, the audiences it addresses, and the constraints (temporal, social, modal, etc.) associated with the genre. We ask students to answer particular questions that prompt them to look more generally at the genre in these terms. They then consider choices they will actually make in working with their chosen genre to communicate their EJ research. Nevertheless, first, we ask them to think about who will help them to make the most informed and relevant genre choice.

For example, we ask students the following questions: Which organizations or individuals with on-the-ground experience might you consider as an **engaged "thinking partner" whom you could consult** with as you make your decision? **Who is the audience** that you want to reach through your EJ research project (and your outputs)? **What communication genres** might help you with reaching these particular groups of people (your chosen audience)?

We then ask students to discuss some examples of the genre they would like **to use** for their project and why this genre is a good fit for communicating their EJ research. Guided questions include:

- Who typically communicates using this genre and what kinds of audiences typically engage with it?
- What purposes does it usually fill for creators and audiences? (These might be the same, or they might diverge.)
- Where and under what conditions might one expect to encounter this genre?
- Finally, what kinds of subjects and themes are commonly addressed (and not addressed) in this genre?
- We also want them to reflect on the formal and stylistic features of the genre, and we ask them to think through the following:
- Are there common structures? Designs? Elements? Moves?
- Are there similar kinds of content included or excluded?
- Are there stylistic patterns? (voice, language, appearance)

Finally, we ask about the available "means of persuasion" afforded by the genre, potential constraints, and what kinds of modal choices they might make. For instance, a research poster advocating for institutional divestment from fossil fuels may be printed and mounted on a poster board, but it may also be projected on a screen. What modality or modalities is the genre typically delivered in? How will the choice of modality impact what they can do (or not do) in this genre and how persuasive they might be in convincing others to take action?

This activity is one of the most important activities of our entire course because it is central to our assessment and grading of the research projects. Once students have created their projects, we ask them to use the questions above to write a comprehensive reflection on their genre choices. We use this reflection to help determine their grade. Colleagues have asked me how we can grade podcasts, digital films, or storymaps if we are not teaching students how to make them? I always respond that our class is less about training students how to make a particular piece of digital media like a podcast. The class is more about getting them to think rigorously about why they would want to make a podcast in the first place and then offering them the space to learn, fail, practice, create, and develop the podcast tentatively and with support from our class community.

While it is true that many students do come with some experience with their chosen genre, many other students want to use our class to experiment and explore. We encourage them to do so by asking them to think carefully and with nuance, not only about the technical production but about the ecosystem in which the digital product is embedded and the potential impact it might have. The same skills they practice in our class—the willingness to take risks, learning something new, making mistakes, pushing themselves to explore out of their comfort zone, working collaboratively, and supporting their peers to create and imagine in ways they haven't before—are the same skills they will need to address the climate crisis. In this way, our attention to the process and the practice is just as important as the product.

Classroom as a Site of Imagination, Exploration, and Hope

It is important to emphasize that the purpose of our course is not only to help students develop the skills to integrate and platform community voices in climate justice digital media projects. It is also a vehicle for creating a nurturing and supportive community grounded in possibility and hope. One cannot effectively train students to intervene in global crises without first teaching them how to take care of themselves and create and sustain communities of care that can do the work of tending to justice—not just outside the classroom but also inside the classroom. My environmental writing classes are an invitation to work collectively to imagine alternatives and solutions to our climate crisis, communicated with intention and shared in a community that often stays close long after our class is over.

This community building begins on the first day of class. Students co-create class norms; add names to a sign-up sheet for bringing snacks to share (during pre-COVID days); participate in small group activities where classmates are encouraged to listen and support each other with generous, kind feedback;

and engage in guided and structured peer reviews of each other's work. The influence students had on each other was always evident in their reflections. For example, my student Evan K. created a multimedia storymap that connected institutional racism and housing segregation to multiple water crises in his home state of Michigan. He had not settled on a genre until he met with a small group to discuss his progress. He wrote in his reflection:

> I was inspired by Sydney's project. She told me about her experience working at a non-native organization as a non-native person recognizing the need to bring in indigenous voices and needs to their river stewardship. I was inspired by her approach—using her positionality as an outsider to gain legitimacy with other outsiders, and then learning as much from the community as possible about the harm the outsiders were causing and how to change. Sydney really helped me hone my audience from just people who generally care, to people who (often unknowingly) perpetuate the cycles I talk about. I was also inspired by Sydney's medium: an interactive article lets you center community voices without over-burdening them and decenters your own voice since there isn't a narrator in the same way as an audio piece. I'm really proud of what I was able to create, and I felt like a lot of the growth that I felt like was slow and subtle ended up being really transformative for me.

Conclusion

Integrating community voices into digital media projects in respectful ways disrupts extractive research cycles, helps make students more sensitive to the important voices not included in climate spaces, and changes narratives that see marginalized communities as victims to platforming them as experts. Over the quarter, students learn how to foment trust with community organizations and each other inside the classroom. They listen deeply, create with courage and care, and experiment with different ways of communicating their research through podcasts, storymaps, data visualization tools, comics, children's books, infographics, digital presentations, and much more. They learn from each other and, perhaps most importantly, figure out what it means to have hope just when they need it the most.

References

Adame, F. (2021). Meaningful collaborations can end "helicopter research." Nature. Retrieved from https://www.nature.com/articles/d41586-021-01795-1

Climate Justice Alliance. Communities taking bold action on the frontlines of climate change. Retrieved August 26, 2022, from https://climatejusticealliance.org/

Fleischman, L., & Franklin, M. (2017). Fumes across the fenceline: The health impacts of air pollution from oil and gas facilities on African American communities. National Association for the Advancement of Colored People, Clean Air Task Force.

Goldman Prize Winner Nalleli Cobo. (2022). Retrieved from https://www.goldman-prize.org/recipient/nalleli-cobo/

Haas Center for Public Service. Principles of Ethical and Effective Service. Retrieved on August 25, 2022, from https://haas.stanford.edu/about/our-approach/principles-ethical-and-effective-service

Hall, B. L. (1992). From margins to center: The development and purpose of participatory action research. *American Sociologist, 23*(4), 15–28.

Haraway, D. (1998). Situated knowledges: The science question in feminism and the privilege of partial perspective. *Feminist Studies, 14,* 575–599. doi:10.2307/3178066

Harding, S. (2004). Introduction: Standpoint theory as a site of political, philosophic, and scientific debate. In S. Harding (Ed.), *The feminist standpoint theory reader: Intellectual and political controversies* (pp. 1–15). Routledge.

Israel, B., Schulz, A., Parker, E., Becker, A., Allen, A., & Guzman, R. (2008). Critical issues in developing and following CBPR principles. In M. Minkler (Ed.), *Community based participatory for health: Process and outcomes* (pp. 47–66). John Wiley & Sons.

Mbaye, R., Gebeyehu, R., Hossmann, S., Mbarga, N., Bih-Neh, E., Eteki, L., & Boum, Y. (2019). Who is telling the story? A systematic review of authorship for infectious disease research conducted in Africa, 1980–2016. *BMJ Global Health, 4*(5). Retrieved from https://gh.bmj.com/content/4/5/e001855

Mikati, I., Benson, A., Luben, T., Sacks, J., & Richmond-Bryant, J. (2018). Disparities in distribution of particulate matter emission sources by race and poverty status. *American Journal of Public Health, 108*(4), 480–485. doi:10.2105/AJPH.2017.304297

Mohai, P., & Saha, R. (2015). Which came first, people or pollution? Assessing the disparate siting and post-siting demographic change hypotheses of environmental injustice. *Environmental Research Letters, 10,* 115008.

North, M. A., Hastie, W. W., & Lauren Hoyer, L. (2020). Out of Africa: The underrepresentation of African authors in high-impact geoscience literature. *Earth-Science Reviews, 208.* doi:10.1016/j.earscirev.2020.103262.

Polk, E., Beach, R., & Webb, A. (2023). Climate crisis literacies in a global perspective. In S. Kerkoff & H. Spires (Eds.), *Critical perspectives on global literacies: Bridging research and practice* (pp. 230–246). Routledge.

Polk, E., & Diver, S. (2020). Situating the scientist: Creating inclusive science communication through equity framing and environmental justice. *Frontiers in Communication, 5*(6). doi:10.3389/fcomm.2020.00006

Pulido, L., & Peña, D. (1998). Environmentalism and positionality: The early pesticide campaign of the United Farm Workers' organizing committee, 1965-71. *Race Gender Class, 6,* 33–50.

Rose, G. (1997). Situating knowledges: Positionality, reflexivities and other tactics. *Progress in Human Geography, 21*(3), 305–320. doi:10.1191/030913297673302122

The Principles of Environmental Justice (EJ). (1991). Environmental Justice/Environmental Racism, EJNet. Retrieved on August 25, 2022, from https://www.ejnet.org/ej/principles.html

Wu, X., Nethery, R. C., Sabath, M. B., Braun, D., & Dominici, F. (2020). Exposure to air pollution and COVID-19 mortality in the United States: A nationwide cross-sectional study. Science Advances. doi:10.1126/sciadv.abd4049

Engaging Students in Imaginative and Critical Thinking through Media Production

As we noted in this book's introduction, students need to reimagine and transform status-quo energy, transportation, agriculture, urban design/housing, and political/legal systems to address the climate crisis. Therefore, the second section of this book focuses on how students employ media productions to critique status-quo systems leading to portraying alternative future actions or scenarios to convince audiences of the need for change.

Engaging students in imaginative/critical thinking through media productions involves helping them learn to consider *how* they produce media to portray climate change impacts for achieving audience uptake leading to changing their attitudes as compared to just sharing images (Isik & Vessel, 2021; O'Neill, 2017). They may use media to portray images of local climate change impacts, for example, the effects of severe weather events on local communities (Scannell & Gifford, 2013) or how the flooding of specific global spaces has an impact on a local community (Terry, 2020). Students may also portray people engaged in adaptation and mitigation practices, such as reducing excessive water use.

It is also the case that producing videos enhances youth's attitudes about the need to take action (Echegoyen-Sanz & Martín-Ezpeleta, 2021). For example, in producing a video about their lives coping with climate change effects on Rigolet Island, Labrador, Canada, Inuit adolescents experienced increased respect from their audiences in ways that enhanced their agency as effective communicators (MacDonald et al., 2015).

DOI: 10.4324/9781003335276-7

Providing Options for Use of Different Digital Media Tools

It is important that students have the opportunity and option to employ a range of different digital media tools and modalities to engage in imaginative/critical thinking, including digital stories, videos, podcasts, art, writing, music, and social media (Milstein et al., 2018). Students may find that they have expertise in or prefer to use certain tools to achieve intended audience uptake. For example, they may create digital story videos to portray images showing themselves engaging in actions related to addressing the climate crisis (Jiang et al., 2019; Smith et al., 2019, 2021). Students may create a digital story portraying characters engaged in planting trees to increase the number of trees absorbing CO_2 emissions. Alternatively, they may begin by writing narratives about themselves, noting instances of climate change effects, leading to generating digital storytelling narratives about the need for more extensive communal actions (Ganz, 2007).

Students can also employ digital visualization tools that portray climate change effects, including photography/art, digital mapping, or infographics *goo.gl/4dgv23*. For example, young people created "Polar Army" artwork *polararmy.org* (Madder, 2017) or art exhibits as evident in the Art for Adaptation project *artforadaptation.com* for using art in ways that change their perceptions of climate change (Bentz, 2020). In addition, they can use visualization tools for acquiring data or images about climate change effects, for example, Climate Interactive *climateinteractive.org*, Climate Scoreboard *climateinteractive.org/tools/scoreboard*, NASA's Global Climate Change *climate.nasa.gov*, Community Viz *placeways.com/communityviz*, National Climate Assessment *nca2014.globalchange.gov*, Visualizing Change Toolkit *goo.gl/J2zgpo*, or Visualizing Change *vischange.org* (Boss, 2019).

For engaging in a critical analysis of causes of emissions, students can also use Google's Environmental Insights Explorer *insights.sustainability.google* to record carbon emissions from buildings and transportation in their cities, finding that large downtown buildings emit a lot of emissions (Meyer, 2018). They may also use the National Institute of Health TOXMAP *toxmap.nlm.nih.gov/toxmap* or EJScreen *epa.gov/ejscreen* to portray power plants' emissions levels near low-income neighborhoods resulting in adverse health effects on those neighborhoods (Beach & Smith, 2020).

Students also employ digital visual images and socio-spatial mapping to portray climate change impacts (Bentz, 2020; Cone et al., 2012; Jocson, 2016). They may use digital apps (for eight apps *t.ly/GPmb*) to generate multimodal posters to showcase their knowledge of renewable energy sources (Castek & Dwyer, 2018). Preservice teachers in a methods course used various media to portray their responses to topics and texts about water justice issues related to water scarcity, privatization, pollution, and ocean warming (Woodard &

Schultz, 2022). In response to texts about water issues, teachers created mind maps portraying connections between factors shaping water justice. Mind maps created at the end of the course demonstrate increased knowledge of water justice issues. A group of teachers also created a zine, "The Book of Hydration: Quenching Thirst from A–Z," that portrayed issues of water conservation, bottled word, hydropower, and water quality challenges (Woodard & Schultz, 2022).

Youth are also creating and performing music to convey their perceptions of climate change (Wodak, 2017), as evident in their engagement with The Climate Music Project *t.ly/vHhF* and The Global Climate Change Music Project YouTube Channel *t.ly/Ea4n*. For example, in a college course, students created videos of their musical performances, leading the instructors to note that

> Clearly the most rewarding aspect of the class is the use of music as a medium for climate change education. Each student is required to provide the class with a YouTube or other music video link from an artist or group regarding climate change, the environment, and/or the need for political and personal responsibility regarding the planet.
>
> *Snow & Snow (2010, p. 64, quoted in Wodak, 2017)*

Youth may employ social media to address climate change, for example, by sharing social media posts on Twitter using #climatecrisis #climatechange (Field, 2021; Napawan et al., 2017), given that 56% of Gen Z youth access topics on climate change on social media and 45% post about climate change on social media (Tyson et al., 2021), as well as how there was an increase in social media focused on climate change topics by more than 100% from 2018 to 2019 (Ellis, 2019).

Youth may employ social media for organizing and promoting demonstrations and strikes, for example, the School Strike 4 Climate protests (Boulianne et al., 2020) associated with engaging in "connective action" mediated through digital interactions (McLean & Fuller, 2016).

A meta-analysis of social media platforms found that Twitter was particularly popular, given that users can share tweets with global audiences using hashtags such as #OurChangingClimate (Pearce et al., 2019). Students used the #OurChangingClimate or #FridaysForFuture, as evident in the climate strike in 2019 that included eight million young people in 150 countries as well as lawsuits initiated by young people in many countries (Napawan et al., 2017). Students at RMIT University in Australia employ Climate Action Adventure! Website to engage in virtual learning by accessing social media platforms to address climate justice issues (Rousell et al., 2021).

At the same time, youth need to think critically about how they respond to and employ social media to acquire and communicate about climate change, given that they often access social media as their primary news source (Anderson & Jiang, 2018; Liu & Kim, 2022). (Of 5,844 college students, 89% accessed news from social media, 76% from online newspapers, and 55% from news feeds (Head et al., 2018).) They may acquire misinformation on climate change topics on social media related to posts that "global warming is a hoax" (Andersson & Öhman, 2016). However, while they often voice alternative perspectives in their interactions, they may not adopt specific critical perspectives, suggesting the need for classroom instructions to apply criteria to ensure sharing and vetting of different perspectives (Thomas et al., 2021).

Students also experience imaginative play through engaging in digital role-play/video games on climate change, such as Climate Quest, World Climate, Fate of The World, ecoKoin, SimCity EDU, Eyes on the Rise, Earth Primer, or The Carbon Cycle (Meadows et al., 2016). For example, in engaging in the Climate Quest game *earthgames.org/games/climatequest* or the World Climate role-play simulation *cleanet.org/resources/43001.html*, students adopt roles for proposing policy positions on addressing climate change effects (Hassall, 2018; Sterman et al., 2015). In addition, in playing the EnerCities *paladinstudios.com/enercities* game, they need to decide on the use of non-renewable versus renewable energy resources in a virtual city (Janakiraman et al., 2021).

Chapter Summaries

In the first chapter of this section, "Our Story Will be the Future: A Learner-centered Approach to Support Digital Multimodal Composing about the Climate Crisis," Shiyan Jiang, Blaine E. Smith, and Ji Shen describe how students engaged in shared, collaborative thinking for selecting to portray particular climate change effects in the Miami area as well as the use of production technique/editing in creating their digital stories. Students participated in the Imagine the Future project at the University of Miami, working collaboratively in teams to create digital stories focused on environmental sustainability issues (Jiang et al., 2019; Smith, 2017, 2019; Smith et al., 2019, 2021; Smith & Shen, 2017).

For creating digital stories, students worked collaboratively in small groups by assuming roles of their choice (e.g., designer, scientist, and writer). They created multilayered digital narratives, including comics, animations, infographics, videos, and images, to propose a creative solution to issues of choices related to climate change. In addition, students engaged in problem-solving choices to portray perceptions of and solutions to climate change impacts. For example, they used problem-solving to portray residential flooding issues in Miami, problem-solving practices that teachers can employ in supporting students' digital storytelling practices.

In the second chapter, "'Listen, There, To the Way the Real World Thinks in Me': Cultivating an Empathic Imagination to Support Students' Visual

Stories that Address the Climate Crisis," Linda Buturian documents how students respond to novels, graphic stories, and poems in her college courses (e.g., *t.ly/XRA7*) to create visual artistic texts. In doing so, her students engaged in imaginative thinking about future climate change scenarios based on students' life experiences, with examples of students' artwork and writing in her chapter. In her courses, she emphasizes the importance of addressing climate justice issues through youth participatory action research. She also posits the need to adopt transdisciplinary methods for teaching about climate change by combining literacy, social studies, and art instructions.

The third chapter in this section, "Addressing Climate Change and Sustainable Energy Futures Through Creative Music Engagement" by Evan S. Tobias, Kyle Bartlett, Michelle E. Jordan, and Steven J. Zuiker, describes how youth create music productions about their perceptions of and attitudes toward climate change. Through their musical productions, students reimagine alternative futures, for example, a future based on the adoption of solar energy. For example, in The Weight of Light (WoL) project, students created music to portray future energy systems based on the use of solar energy. The chapter also draws on the United Nations Educational Scientific and Cultural Organizations (UNESCO) framework of Education for Sustainable Development (ESD) to posit the value of music educators for engaging students in creating and performing music as a medium for reimagining future worlds.

The final chapter in this section, "Fostering Proactive Ecological Identity of Youth Through Social Media" by Nataliia Goshylyk, describes young people's use of social media for interacting about climate change. In using social media, youth identified instances of misinformation about climate change by employing three different types of practices:

1 Learning to comprehend and assess ecological information on social media.
2 Analyzing ecological information within the global communication framework.
3 Setting local ecological agenda using social media as an interaction platform leading to engagement in ecological community concerns.

The chapter draws on Bloom's Digital Taxonomy to identify how students progress through different activities to analyze ecological challenges in their communities based on case studies of youth engagement in European projects of students using linguistic devices, multimedia resources, phrasal connections, and hashtags to interact with others.

For additional links, resources, and readings for this section, refer to the book's website *t.ly/S3X2*.

References

Anderson, M., & Jiang, J. (2018). Teens, social media & technology 2018. *Pew Research Center, 31*, 1673–1689. https://www.pewresearch.org/internet/2018/05/31/teens-social-media-technology-2018/

Andersson, E., & Öhman, J. (2016). Young people's conversations about environmental and sustainability issues in social media. *Environmental Education Research, 23*(4), 465–485. doi:10.1080/13504622.2016.1149551

Beach, R., & Smith, B. E. (2020). Using digital tools for studying about and addressing climate change. In P. M. Sullivan, J. L. Lantz, & B. A. Sullivan (Eds.), *Handbook of research on integrating digital technology with literacy pedagogies* (pp. 346–370). IGI Global.

Bentz, J. (2020). Learning about climate change in, with and through art. *Climatic Change, 162*, 1595–1612. doi:10.1007/s10584-020-02804-4

Boss, S. (2019, January 11). Teaching climate change across subjects. [Web log post]. https://goo.gl/xNsa9e

Boulianne, S., Lalancette, M., & Ilkiw, D. (2020). "School Strike 4 Climate": Social media and the international youth protest on climate change. *Media and Communication, 8*(2). doi:10.17645/mac.v8i2.2768

Castek, J., & Dwyer, B. (2018). Think globally, act locally: Teaching climate change through digital inquiry. *The Reading Teacher, 71*(6), 756–761. doi:10.1002/trtr.1687

Cone, J., Rowe, S., Borberg, J., & Goodwin, B. (2012). Community planning for climate change: Visible thinking tools facilitate shared understanding. *Journal of Community Engagement and Scholarship, 5*(2), 7–19.

Echegoyen-Sanz, Y., & Martín-Ezpeleta, A. (2021). Fostering creativity in the classroom: Ecofeminist movies for a better future. *Journal of Education Culture and Society, 1*, 117–130. doi:10.15503/jecs2021.1.117.130

Ellis, K. K. (2019, September 26). How social media is driving the climate change conversation. NewsWhip. http://t.ly/MCkf

Field, E. (2021). Is it all just emojis and lol: Or can social media foster environmental learning and activism? In M. Hoechsmann, P. R. Carr, & G. Thesee (Eds.), *Education for democracy 2.0: Changing frames of media literacy*. Brill/Sense Publishers.

Ganz, M. (2007). *Worksheet: Telling your public story: Self, us, now.* www.welcomingrefugees.org/sites/default/files/documents/resources/Public%20Story%20Worksheet07Ganz.pdf

Hassall, L. (2018). Climate literacy and collaborative on-line landscapes: Engaging the climate conversation through drama facilitation in distance and e-learning environments. In U. M. Azeiteiro, W. L. Filho, & L. Aires (Eds.), *Climate literacy and innovations in climate change education* (pp. 375–388). Springer.

Head, A. J., Wihbey, J., P., Metaxas, P. T., MacMillan, M., & Cohen, D. (2018). How students engage with news: Five takeaways for educators, journalists, and librarians. Project Information Literacy Research Institute. http://www.projectinfolit.org/uploads/2/7/5/4/27541717/newsreport.pdf

Isik, A. I., & Vessel, E. A. (2021). From visual perception to aesthetic appeal: Brain responses to aesthetically appealing natural landscape movies. *Frontiers in Human Neuroscience, 15*, 676032. doi:10.3389/fnhum.2021.676032

Janakiraman, S., Watson, S. L., Watson, W. R., & Newby, T. (2021). Effectiveness of digital games in producing environmentally friendly attitudes and behaviors: A mixed-methods study. *Computers & Education, 160*, 104043. doi:10.1016/j.compedu.2020.104043

Jiang, S., Smith, B. E., & Shen, J. (2019). Examining how different modes mediate adolescents' interactions during their collaborative multimodal composing processes. *Interactive Learning Environments, 29*(5). doi:10.1080/10494820.2019.1612450

Jocson, K. M. (2016). "Put us on the map": Place-based media production and critical inquiry in CTE. *International Journal of Qualitative Studies in Education, 29*(10), 1269–1286.

Liu, B. F., & Kim, J. (2022). Social media and climate change dialogue: A review of the research and guidance for science communicators. In J.-W. Yusuf & B. St. John (Eds.), *Communicating climate change: Making environmental messaging accessible* (pp. 97–115). Routledge.

MacDonald, J. P., Ford, J., Willox, A. C., & Mitchell, C. (2015). Youth-led participatory video as a strategy to enhance Inuit youth adaptive capacities for dealing with climate change. *Arctic, 68*(4), 486–499.

Madder, B. (2017, April 3). *Kids' 'Polar Army' artwork shows youths' concerns* [Web log post]. https://goo.gl/k9VREW

McLean, J. E., & Fuller, S. (2016). Action with(out) activism: Understanding digital climate change action. *The International Journal of Sociology and Social Policy, 36*(9/10), 578–595. doi:10.1108/IJSSP-12-2015-0136

Meadows, D., Sweeney, L. B., & Mehers, G. M. (2016). The climate change playbook: 22 *Systems*-thinking games for more effective communication about climate change. Chelsea Green Publishing.

Meyer, R. (2018, September 25). *Google's new tool to fight climate change* [Web log post]. https://goo.gl/4VjKzN

Milstein, M., Pileggi, E., & Morgan (Eds.). (2018). *Environmental communication: Pedagogy and practice.* Routledge.

Napawan, N. C., Simpson, S.-A., & Snyder, B. (2017). Engaging youth in climate resilience planning with social media: Lessons. *Urban Planning, 2*(4), 51–63. doi:10.17645/up.v2i4.1010

O'Neill, S. (2017). Engaging with climate change imagery. *Oxford Research Encyclopedia of Climate Science.* doi:10.1093/acrefore/9780190228620.013.371

Pearce, W., Niederer, S., Özkula, S. M., & Querubín, N. S. (2019). The social media life of climate change: Platforms, publics, and future imaginaries. *WIREs Climate Change, 10*(2), e569. doi:10.1002/wcc.569

Rousell, D. (2021). Cosmopolitical encounters in environmental education: Becoming-ecological in the intertidal zones of Bundjalung National Park. *The Journal of Environmental Education, 52*(2), 133–148. doi:10.1080/00958964.2020.1863313

Scannell, L., & Gifford, R. (2013). Personally relevant climate change: The role of place attachment and local versus global message framing in engagement. *Environment and Behavior, 45*(1), 60–85. doi:10.1177/0013916511421196

Smith, B. E. (2017). Composing across modes: A comparative analysis of adolescents' multimodal composing processes. *Learning, Media & Technology, 42*(3), 259–278.

Smith, B. E. (2019). Mediational modalities: Adolescents collaboratively interpreting literature through digital multimodal composing. *Research in the Teaching of English, 53*(3), 197–222.

Smith, B. E., Beach, R., & Shen, J. (2021). Fostering student activism about the climate crisis through digital multimodal narratives. *Journal of Sustainability Education.* http://t.ly/sXlV

Smith, B. E., & Shen, J. (2017). Scaffolding digital literacies for disciplinary learning: Adolescents collaboratively composing multimodal science fictions. *Journal of Adolescent & Adult Literacy, 61*(1), 85–90. doi:10.1002/jaal.660

Smith, B. E., Shen, J., & Jiang, S. (2019). The science of storytelling: Middle schoolers engaging with socioscientific issues through multimodal science fictions. *Voices from the Middle, 26*(4), 50–55.

Snow, R., & Snow, M. (2010). The challenge of climate change in the classroom. *WSEAS Transactions on Environment and Development, 6*(1), 74–83.

Sterman, J., Franck, T., Fiddaman, T., Jones, A., McCauley, S., Rice, P., Sawin, E., Siegel, L., & Rooney-Varga, J. N. (2015). World climate: A role-play simulation of climate negotiations. *Simulation & Gaming, 46*(3–4), 348–382. doi:10.1177/1046878113514935

Terry, M. (2020). *The geo-doc: Geomedia, documentary film, and social change.* Palgrave.

Thomas, L., Hernandez, I., & King, A. V. (2021). These social media activists prove digital learning isn't just zoom class. *Green America, 18*, 19.

Tyson, A., Kennedy, B., & Funk, C. (2021). Gen Z, millennials stand out for climate change activism, social media engagement with issues. Pew Research Center. https://www.pewresearch.org/science/2021/05/26/gen-z-millennials-stand-out-for-climate-change-activism-social-media-engagement-with-issue/

Wodak, J. (2017). Shifting baselines: Conveying climate change in popular music. *Environmental Communication.* http://t.ly/Uwo8D

Woodard, R., & Schultz, K. (2022). Developing urgent writing pedagogies in teacher education: Portraits of practice from an inquiry into water justice. In T. S. Hodges (Ed.), *Handbook of research on teacher practices for diverse writing instruction* (pp. 270–295). IGI Global. doi:10.4018/978-1-6684-6213-3.ch014

5

OUR STORY WILL BE THE FUTURE

A Learner-centered Approach to Support Digital
Multimodal Composing about the Climate Crisis

Shiyan Jiang, Blaine E. Smith, and Ji Shen

Introduction

Many socioscientific issues (SSIs) arise as our society advances at an unprecedented pace through anthropocentric development. In a nutshell, SSIs refer to social issues and dilemmas related to science and technology development, such as environmental issues related to biotechnology applications (Sadler, 2011). Since discourse and policies about these issues have penetrated our daily conversations and popular media (Sadler & Donnelly, 2006), making evidence-based arguments and informed decisions on SSIs has become an important part of being scientifically literate (Zeidler et al., 2002). Therefore, scientific literacy should be a goal for all students regardless of their future career trajectories (e.g., in STEM or not). Learning about SSIs has demonstrated an effective way of helping students develop scientific literacy (Sadler, 2004; Sadler & Donnelly, 2006; Sadler & Zeidler, 2009).

Climate change is one of the pressing SSIs that impacts the world and its inhabitants. Students need to be aware of climate change and its effects, and they need to be equipped with the knowledge and tools to take action to mitigate and adapt to climate change (Eide & Kunelius, 2021; Karsgaard & Davidson, 2023). Efforts have been devoted to increasing youth literacy on climate science, including through the development of curriculum and resources (Busch et al., 2019; Busch & Ayala Chávez, 2022; Smith et al., 2021). However, there is still a lack of research on strategies for empowering student voices, helping students to connect SSIs to their personal life, and guiding students to infuse their innovative thinking and solutions to SSIs in their digital products that they can share with the rest of the world.

DOI: 10.4324/9781003335276-8

This chapter describes a learner-centered approach to support multimodal composing (Jewitt, 2008; Kress, 2003) to address the climate crisis. Specifically, we first describe Project Imagine the Future (IF), focusing on the design of multiple elements of choices (i.e., disciplinary roles, modes of expression, and climate change indicators). These choices allowed students to connect the climate crisis to their personal life and to share their innovative thinking and solutions to the climate crisis through digital multimodal composition. We then illustrate how students navigated these choices and discuss implications for practices when adopting this approach in various learning contexts.

Project IF: A Learner-centered Multimodal Composing Environment

Project IF is a design-based research study (Brown, 1992; Collins, 1992) that has evolved through five implementations over three years. The duration for each implementation varied based on the context of implementation (Mean = 38.9 hours; SD = 22.8). We tailored the implementation time to fit students' needs in different contexts, including an after-school program, a selective STEAM course, and a hybrid informal program. Throughout a three-year study, 130 students from diverse backgrounds (11, 9, 32, 42, and 36 students in the first, second, third, fourth, and fifth implementation, respectively) formed 35 groups to create multimodal science fiction stories. In science fiction stories, students were required to propose a creative solution to climate change and human health issues.

Insights from these implementations were used to modify the intervention design iteratively. In particular, in the iterative design, we developed a scaffolded learner-centered approach to empower student voices by offering multiple elements of choice. In this section, we describe one design feature of the approach, identity-conscious choices (i.e., disciplinary roles, modes of expression, and climate change indicators). We define identity-conscious choices as a set of choices that acknowledges students' diverse identities, interests, and backgrounds. Offering identity-conscious choices is critical for students to feel engaged with learning activities. Doing so empowers them to express themselves and encourage them to see themselves as experts in multimodal composition and as agents who can take actions to improve their communities' environmental sustainability.

Disciplinary Roles

In the project, students selected disciplinary roles of their choice and worked collaboratively in small groups of three to five to create multimodal science fiction stories. In the first implementation, students self-selected a role among four given ones (i.e., artist, engineer, scientist, and writer). Artists led the creation of visual and audio representations for main characters and/or scenes in the story. Engineers were responsible for designing the story's buildings, vehicles, and settings. Scientists were in charge of verifying and incorporating scientific information related to the story. Finally, writers were accountable for developing and writing the fiction plot. Based on classroom observations and interviews, we found that students who took the engineer role either did the same thing as artists or left without knowing what to do (Jiang et al., 2019). Thus, we combined the roles of artist and engineer into one role, designer, in later implementations. Designers were responsible for creating visual and audio representations (e.g., comics and animations) related to the story.

Students were encouraged to negotiate their roles and responsibilities with group members. Our work demonstrated that students were likelier to participate in the composition process and demonstrate ownership of their work when they were engaged in making decisions about their roles (Jiang, 2023). Furthermore, as these roles are discipline-specific, changing roles could provide students with opportunities to develop multiple discipline-specific practices and understand the connection between different disciplines. For example, although students self-selected to be a writer at the beginning, they could still contribute to the design of animation as a designer while being a writer as a major role (we call it hybrid roles; Jiang et al., 2020). They might even eventually change their roles to be designers based on team negotiation. In this way, students could collaborate to develop the story, experience writing, and design practices, and identify connections between writing and design.

However, we acknowledge that role-changing is not always necessary or desirable, and some students may wish to maintain the same role throughout the process. In such a manner, students can focus on developing practices in a particular discipline and continue building on their competencies in that area. Nevertheless, balancing the need for students to develop in-depth disciplinary practices while providing opportunities to explore different practices and understand how disciplines are connected through disciplinary role-taking is a complex task and requires careful planning and attention.

Modes of Expression

Students utilized various digital tools to create multilayered and multimodal digital narratives, such as comics, animations, infographics, and videos (Jewitt, 2008; Kress, 2003; Smith et al., 2019). We carefully selected tools and modes, designed activities in which students could bring their life experiences, and created spaces for students to form learning communities. For example, students learned to create multimodal comics with Pixton (Azman et al., 2015) to show their future professions and working environments. These comics were then shared in the Pixton community, where others could comment on and remix them. After learning these tools to create modes individually (e.g., comics to show future professions), students discussed how to use the modes to support the development of multimodal science fiction stories in group projects. Next, students chose tools and modes they were interested in or comfortable with and contributed to group work by creating corresponding modes. Beyond tools and modes, we introduced, students were encouraged to bring in multimodal tools that they were familiar with, and we provided technological support.

Modes and some disciplinary roles are intertwined. For instance, a student who took the role of writer might be more likely to create textual modes, such as writing a story. However, the student might also be interested in creating a video to support the story. In this case, the student took a hybrid role of writer and designer. The intertwined relationship made students aware of roles that group members took and change roles flexibly. When students were aware of group members' roles, they could utilize each other's expertise and support each other (Jiang et al., 2019). Unlike the roles of writer and designer, scientists could write texts, create videos, or choose other modes to explain scientific ideas. These modes could support students in developing disciplinary identities (e.g., science identity) and cross-disciplinary boundaries (Jiang et al., 2021). For example, when scientists became familiar with creating videos to explain science ideas, they could also use videos to introduce the story's background as a designer. Overall, modality choices provide venues for students to become familiar with disciplinary practices and to cross disciplinary boundaries.

In the iterative design of modality choices, we considered how to support the meaningful integration of multiple modes through technology development. We started by guiding students to create multimodal science fiction stories as a webpage and ended up supporting students in designing interactive books (Figure 5.1) using multimodal artifacts created from different platforms

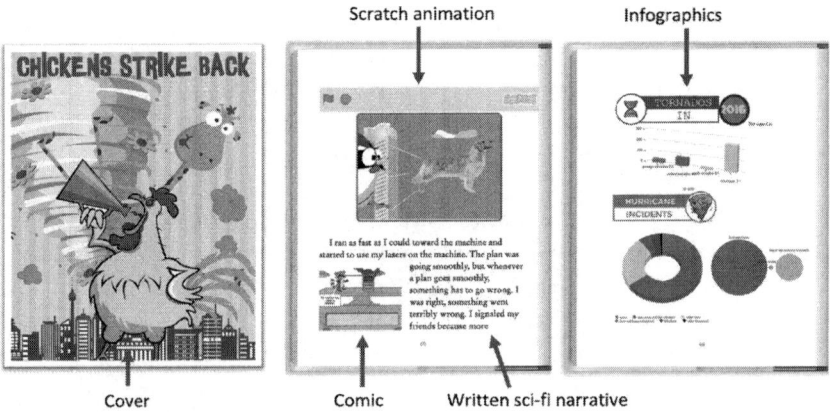

FIGURE 5.1   Examples of Student-created Multimodal Science Fiction Stories, in the Format of Interactive Books

in iKOS (a multimodal composing platform; Shen et al., 2020). For some group projects, we turned interactive books into physical books in which dynamic modes, such as videos and animations, were replaced with QR codes for readers to scan and view them. Book design allowed students to create a tangible product they could share with others and have an intimate relationship with, such as viewing themselves as book authors and editors.

Climate Change Indicators

In science fiction stories, students were required to propose a creative solution to issues of choices related to climate change. These issues are climate change indicators we might observe or experience in the world around us. Moreover, students were challenged to use their imagination to develop story ideas related to climate change indicators that were both creative and scientifically sound. To support the development of scientifically credible stories, we provided science-related activities such as learning science concepts in WISE (Linn et al., 2003), attending lectures presented by guest speakers from different science backgrounds, and visiting science labs at a local university. These activities were designed to broaden students' exposure to climate change indicators and provide them with opportunities to connect science to real-world problems.

In addition, we chose climate change as it is a locally relevant science theme. Students in Project IF were living in an area particularly vulnerable to climate change's impacts. Rising sea levels due to climate change are a major concern for the city where students live. Also, hurricanes and other tropical storms seriously threaten coastal communities such as the one in which students lived. As a result, students could connect the theme of climate change to their personal experiences and the local community. Locally relevant science themes also helped students to generate realistic and concrete narratives for their stories. For instance, their experiences of preparing for hurricane season could be turned into rich and vivid story plots that the characters in their stories could experience. Furthermore, we considered a common theme as we would be well positioned to provide support in science learning and thus guide students to gain an in-depth view of climate change indicators that they chose.

In summary, choices in disciplinary roles, modes of expression, and climate change indicators provided compositional flexibility. This compositional and creative freedom can potentially encourage, engage, and empower student voices and support the free expression of perspectives toward addressing the climate crisis and how it connects to their current and future lives. In the following section, we present examples of how students navigated these choices to describe how the navigation contributed to their understanding of climate change indicators, the impact of such changes, and ways of addressing the climate crisis. When presenting student examples, we draw evidence from pre-, mid-, and post-surveys conducted to gather student perspectives on their learning experiences on the first day, in the middle, and on the last day of project implementation (Sfard & Prusak, 2005). In addition, we present evidence analyzed from student artifacts and end-of-project interviews (Benwell & Stokoe, 2016).

Students' Navigation of Disciplinary Role, Modality, and Climate Change Indicator Choices

Within their groups, students adopted different roles and employed different modes, as choices constituting their digital story productions.

Disciplinary Role Choices: Role-taking and Changing

Most students selected roles based on disciplinary interests and strengths. Through disciplinary role-taking, they led different aspects of the group work. For example, this group work might contribute to distributed leadership in the group and distributed ownership of multimodal artifacts (Jiang et al., 2021). Moreover, some students changed roles and engaged in multiple disciplinary practices through role changing. As a result, they built a shared understanding of the group project, learned the connections between disciplines, and extended their comfort zone of practice by performing unfamiliar practices.

For instance, a team of three male students (Nick, Alex, and Brandon; all names are pseudonyms; six graders) created a story called "What Would Happen if the World Stopped Spinning" (Figure 5.2). The story did not address the

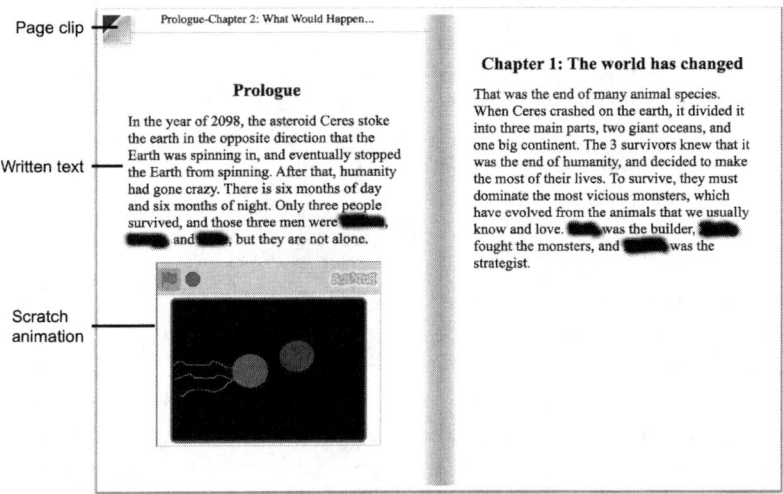

FIGURE 5.2 A Screenshot of a Multimodal Science Fiction Story, "What Would Happen if the World Stopped Spinning"

climate crisis directly; instead, it presented a radical scenario of environmental change for readers to reflect on human dependency on the environment. In the story, three characters (Nick, Alex, and Brandon used their names for characters; each character represented one team member) survived after an asteroid struck the Earth. The strike of asteroid Ceres caused the stop of Earth spinning. As a result, there was no alternation between day and night, and the Earth has become a deserted place with radiation. The three characters found each other, built a shelter, and fought mutated monsters together.

Alex started the idea of the Earth stopping rotation as he was always curious about how the universe worked and what would happen if the environment changed substantially. He pointed out, "I explained many realistic theories if the world stopped spinning" (post-survey). His interest in cosmology drove the background of the story. After agreeing on the background proposed by Alex, the group developed the plot together, and each led the narration and description of their character's experiences. In developing the story, the team brought their story ideas, negotiated, and modified their ideas to create a coherent multimodal science fiction story. Their final story includes three comics (Nick's contribution as a designer), one image (Nick's contribution as a designer), two animations (Alex's contribution as a scientist), and 1,177 words (Brandon's and Nick's contribution as writers).

In this group, Alex (scientist) and Brandon (writer) selected roles they were interested in and perceived as experts. However, Nick chose the role of designer as he was most confident in design practices; he was most interested in the role of a scientist. As he reported, "I would like to be a designer because I can design how things look" (pre-survey). In contrast, although he "strongly liked" science, he was "undecided" about being a scientist in the team. He explained, "I was not sure if I will be good at the scientist role" (pre-survey). In addition, he "disagreed" with being a writer because he didn't like writing as he "can't think of anything to write about" (pre-survey). These responses suggested that Nick was clear about his strengths and selected a role in which he had the confidence to perform associated practices.

Taking on the role of the designer, Nick became more interested in the role of the designer over time. First, he had confidence in designing practices and could design visuals. Second, being able to perform design practices not only built his confidence in being a designer but also facilitated his understanding of the connections between design and writing. In this way, he built confidence in writing, a practice with which he was less comfortable, which might contribute to his increased interest in taking the designer role.

One of the most interesting findings about Nick's disciplinary role-taking was that he gradually built confidence in writing, developed interest in the role of writing, and took the hybrid role of designer and writer. He admitted that he "hated writing" in the pre-survey but said he liked writing in the other two surveys. In the mid-survey, he responded, "I like to be a writer because I am confident that I can create the writing, and I like to write stories" (mid-survey). While responding to the question "what's your favorite part of the project" in the final interview, he stated, "the writing was cool. I liked how we got to write our own book." His responses indicated that he not only became confident in performing writing practices but also developed a strong interest in writing.

Nick's changing attitudes toward writing and being a writer were consistent with his changing attitude toward taking roles in general. Initially, he "agreed" that he preferred taking roles but did not value them, "I do not care whether I get a specific role or not because I know I will enjoy whatever I get" (pre-survey). In the mid-survey, he still "agreed" with his preference for role-taking after trying out different roles, "I like to take on designer and writer because I am not sure if I want to be a scientist." The responses suggested that role-taking provided opportunities for Nick to carry out diverse disciplinary practices and develop diverse perspectives. In the end, he "strongly agreed" that he would like to take roles, "I do prefer taking on a specific role. While I'm good at taking on the roles that I have now, I'm not very good at the other role" (final survey). From his perspective, role-taking could strengthen specific skills and knowledge.

The attitude change in role-taking was mediated by the perception of how role-taking should work. He perceived role-taking as a way to participate in different disciplinary practices. "I kind of tried out writer. I tried out pretty much all three roles. I just wasn't the best at scientist and I kind of like writing on the computer. I never really tried it before" (final interview). He further emphasized the power of role-taking in enabling him to try out things, including those he felt less confident with, "if I had every single role, I would be all over the place I would think. I don't really know what I should do; I don't know what I'm best at. I do." Through role-taking, he could reach beyond the comfort zone of design and not only tried out writing but also developed a strong interest and competence in writing.

His attitude toward taking the scientist role stayed neutral even though he had a strong interest in science throughout the project. His responses to the question of interest in science and preference for being a scientist in the team were the same in the three surveys. On the one hand, he "strongly agreed" that he liked science and wanted to explore any science topics. On the other hand, he was "undecided" about being a scientist in the team because he was unsure whether he could do it well. The fact was that the scientist of his team, Alex, demonstrated very strong scientific expertise and proposed the driving science topic (i.e., the Earth stopped spinning) for their group's fiction. Alex was also very outspoken as the team's scientist, as he was expressive in explaining science ideas during group discussions, whole class discussions, and (especially) presentations. Moreover, Alex preferred working by himself. The presence of a strong scientist who preferred working alone might prevent Nick's more active participation as a scientist in the team.

Like many other students in the project, Nick chose roles that he perceived as an expert. By choosing a role they feel they are an expert in, students might be more likely to feel empowered to contribute to the group project and the class as a learning community. Additionally, like Nick, many students changed roles or desired to change roles. However, role changing is not an individual decision but rather a process negotiated among group members. Other groups might decline some students' role changing, and group members might even compete for a specific role. For example, all group members might want to take the designer role as they were interested in making multimodal comics after learning the tool. Thus, group dynamics might influence students' role selection and experience of role changing.

Modality Choices: Expressing Identities

We observed that students had their modal preferences (Smith, 2017). For example, a scientist can use animations to illustrate the striking of an asteroid on the Earth, videos to explain the consequences of extreme weather, and multimodal comics to show the formation of a tornado. Students' technology experience and confidence played a role in their mode preferences (e.g., students who had little confidence in programming might not choose to create animations with Scratch, a block-based programming language; Resnick et al., 2009). Instead, they used preferred modes to show technology expertise and express their identities (e.g., career aspirations, backgrounds, and interests).

As mentioned in the previous section, Alex took the role of scientist and created two science animations for the story. The two animations that Alex created required advanced coding skills in Scratch. Also, he created a Scratch studio in which he was the manager, showing his emergent idea of building a community for science animations. In addition, he added his animations to the class repository, which indicated his willingness to share knowledge with the community. Finally, in the mid-survey, he explained why to choose Scratch, "(I am into) math, engineering, and technology. Using code requires these." His response implied that he viewed coding as an important skill in STEM.

One of the two animations was inserted after the story's prologue (Figure 5.2). It depicted the striking of an asteroid on the Earth. The animation was used to visualize texts that precede it: "In the year of 2098, the asteroid Ceres struck the Earth in the opposite direction that the Earth was spinning in, and eventually stopped the Earth from spinning" (student artifact, multimodal science fiction story). The other animation was displayed at the end of the story. It showed a rocket flying to the Earth (Figure 5.3). This animation was to preview that the story would continue with new visitors to the Earth.

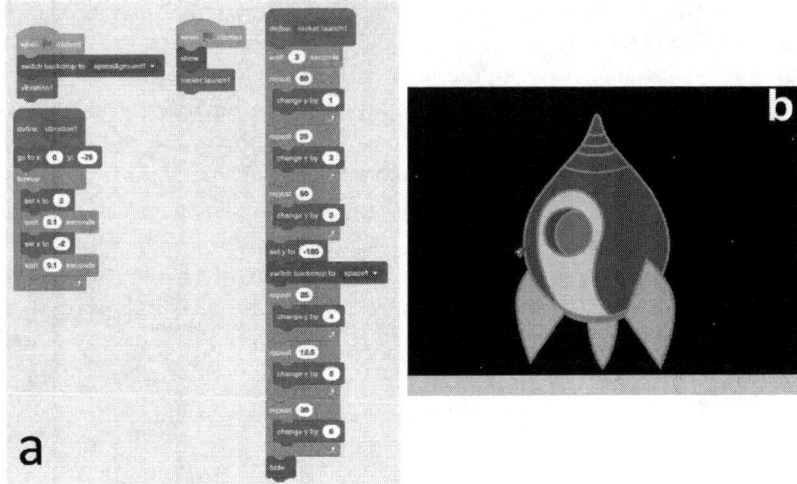

FIGURE 5.3   The Animation of Rocket Flying to the Earth (b) and Coding behind the Animation (a)

The animation of the Earth being struck by asteroid Ceres reflected his current interest in understanding how the universe worked. Throughout the project, he expressed an interest in cosmology. In the interview, he shared, "I really find science fascinating and I love learning about how our world works." His interest in exploring the universe was also related to his career preference of being a rocket scientist or engineer.

Alex's preferred future profession was rocket scientist or engineer throughout the project. The animation of a rocket flying to the Earth reflected his preferred future career. In the "Me in 20 Years" activity, Alex designed a three-panel comic. The comic shows that he found a spacesuit and created a space civilization. Consistent with the comic role, the surveys explicitly showed his strong interest in having an STEM career, particularly in becoming a rocket scientist or engineer. While responding to the item "rate the following statement and explain your rating: my future career/job will be closely related to STEM," his answers in the pre-, mid-, and post-surveys were as follows: "Strongly agree. A rocket engineer," "Agree. When I grow up, I'll be a rocket engineer," and "Strongly agree. A rocket scientist, explosives engineer, or a robotics engineer." The interview further confirmed his career interest. When asked about whether the project helped him to think about future careers, he noted, "Yeah, actually a little bit. I'm thinking I might grow up to be some kind of rocket engineer and explore space someday." He persistently chose a rocket scientist or engineer as a preferred future career and used preferred modes to illustrate career aspirations.

We presented Alex's case as an example of how students created multimodal science fiction stories with their preferred modes of expressing identities. Composing with multiple modes might provide opportunities for students to learn about science and technology in an engaging and meaningful way. This case also showed that technology experience and confidence might influence students' mode preferences. In the project, only a few students chose Scratch to create animations. It might be because Scratch requires coding skills, and some students lacked confidence in coding. While there is a learning curve for students who are new to coding, coding allows students to create interactive and sophisticated animations. Thus, it would be beneficial to provide technology support to help students overcome this obstacle.

Climate Change Indicator Choices: Extreme Weather

One common climate change indicator that students incorporated in multimodal science fiction stories was extreme weather (e.g., hurricanes and floods). The city where students lived was often affected by extreme weather. They had witnessed the severe weather events and their impact on their communities and the city.

As an example, a team of four female students (Sunny, Emma, Jennifer, and Melani), all sixth graders, wrote a story called "Chickens Strike Back" (Figure 5.1). The story has a prologue and 15 chapters. It includes 50 comics, 2 videos, 1 animation, 1 infographic, and 5,180 words. In the story, four animals, each representing one team member and each having its unique superpower, saved New York City as people in the city were endangered after chickens made a machine that caused a tornado.

When asked where the story ideas came from and how the group worked together to make them happen, Sunny shared, "It started by, we wanted natural disasters to deal with, and we also wanted to make it somewhere on Earth. So, the favorite place that we decided on was New York." Emma further clarified, "we got New York because Sunny, she started looking up pictures of tornadoes and hurricanes, and most pictures were in New York. So, we all agreed on New York." Emma's response indicates that the group did not question the scientific accuracy of a tornado or hurricane striking New York. Sunny added, "So we kind of had that idea of you know, New York is very, mostly, the popular place to make a movie or something." As we can see, the group carefully considered the story's details, including the best possible location, and needed support with the scientific aspects of the story.

Often, students did not have a clear justification for the formation of extreme weather in the narrative. Some of them did it intentionally to make the story humorous and at the same time, incorporate the science of climate change. Using the same group as an example, Sunny explained:

> We wanted to make it [the story] funny, and you know, that disaster, and we came up with the chicken. And we came up with the name because of McDonald's. So, we called him McChicken. And so, we wanted to make him evil [creating tornado] because we had this idea of having these superheroes [representing themselves] defeating the chicken.
>
> *Group interview*

Sunny's response indicates that although the group was not explicit in their justification of the extreme weather in the story, they intentionally added humorous aspects to the story (in this case, chickens created a machine to cause a tornado). This does not mean that students did not have a solid understanding of extreme weather and its connection to climate change. This group spent considerable time searching for resources to explore the science behind extreme weather and the global climate crisis. The goals of writing an entertaining story and incorporating scientific ideas were two competing goals in this group's story. Consequently, the group placed more emphasis on the former goal and opted for a less scientific explanation of the formation of extreme weather. This is consistent with our observation of other groups' composing processes. While students' multimodal science fiction stories might lack sufficient scientific explanations, they were able to clearly explain the scientific connection in the group interviews.

While some students were not interested in providing scientific explanations in the story, they provided rich scientific descriptions of extreme weather and the consequence of such weather. In addition, their knowledge of the science of climate change has increased their awareness of the potential impact of climate change on their community. For example, when composing the story "Chickens Strike Back," Sunny searched online sources and shared that she found that tornados were often accompanied by lightning as wind results in the collision of air and ice particles, causing charged clouds. Then lightning is formed by the attraction between the negatively and positively charged clouds. Thus, in the story, the group created four chapters in which the main characters evacuated people in New York City from building getting on fire due to lightning. While the science concepts were not explicitly explained, this group seamlessly integrated them into the narrative (as shown in Figure 5.4).

Sunny's group, like many other groups in the project, chose extreme weather as a climate change indicator. Their lived experiences might inform the choice of this indicator as they had witnessed extreme weather events, including hurricanes and floods, in their communities and cities. The students could connect their personal experiences to the larger issue of climate change. In many cases,

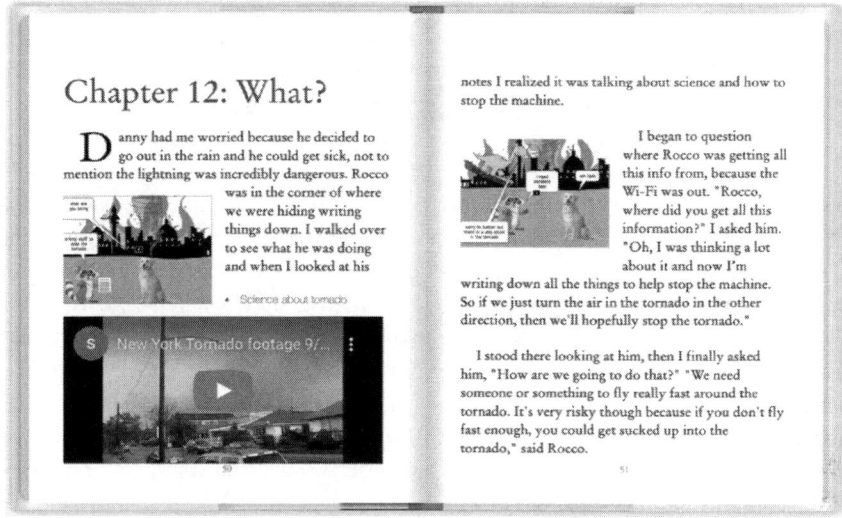

FIGURE 5.4 A Screenshot of a Multimodal Science Fiction Story, "Chickens Strike Back" (Chapter 12)

they have probably seen the destruction that extreme weather can cause in their communities. By writing about extreme weather, they brought attention to this important issue and highlighted its effects on everyday life.

Students portrayed other climate change indicators in their stories, including rising sea levels; shortages of oxygen, water, and food; and endangered aquatic species. To address the climate crisis, some students proposed technological solutions, such as developing renewable energy sources or inventing a machine that can generate oxygen. Other students proposed social solutions, such as creating awareness about climate change or working together to reduce greenhouse gas emissions. Overall, the students' stories demonstrated a good understanding of the science behind climate change and the potential solutions to the problem. The solutions proposed by students might not be feasible in the near future, but they offer a glimpse into the students' scientific understanding and imagination.

Implications for Practice

The project suggests that disciplinary role-taking held the promise of creating a collaborative learning environment in which students could develop expertise in certain disciplines and identify connections between disciplines. Also, it shows that students had mode preferences in multimodal composition and used these modes to express their own identities. Furthermore, students' choices of climate change indicators in multimodal science fiction stories were informed by their lived experiences. This project has implications for implementing a learner-centered approach to teaching climate change and providing scaffolding strategies for adapting similar projects in various classroom contexts.

First, when designing roles, we suggest that educators be mindful of disciplinary boundaries and allow for distributed leadership and ownership to emerge. Meanwhile, we should help students see the connections between disciplines. In addition to supporting role changing for students to understand connections between disciplines, another strategy is to ask students to set composing goals in each group work session. In Project IF, when setting goals, students were required to clearly state artifacts they would create and the help they would need from group members. In such a way, students can make connections between disciplines by reflecting on team members' roles. Overall, disciplinary role-taking was a powerful tool for students to develop expertise in certain disciplines, identify connections between disciplines, and be aware of team members' expertise and needs.

Second, educators should provide freedom for students to move across disciplinary practices, such as using flexible role-taking. Role-taking should be utilized to try out different practices while excelling in one aspect simultaneously, instead of being viewed as a rule one had to follow (some students in earlier implementations had this perception toward role-taking in Project IF). In addition, educators should stress that role-taking is a way to help others in different disciplinary practices. Furthermore, educators should identify those who stayed in roles they did not like but could not switch roles due to a different perception of role-taking or group dynamics and provide corresponding support. Role shuffling could be a strategy to help students re-direct their understanding of (inter) disciplinary practices and roles.

Third, the role-taking strategy negatively and positively impacts student learning while team members are absent. The absence of team members is a critical challenge for informal learning environments. There need to be strategies to deal with missing members to ensure that they will not have negative impacts. As an example, when there is only one student left in a group, educators can guide the student (1) to join other teams while sustaining the same role or changing roles or (2) to invite members from other groups to cover missing roles. Also, missing members may come back after a long time of absence. In this scenario, educators can ask each group to keep track of the group progress in notebooks so that missing members can get familiar with the group project easily. Regarding positive effects, students can participate in other disciplinary practices when missing members are missing.

Fourth, educators should provide modal flexibilities (Smith, 2017) so that students can use tools and modes through which they can express their identities. As students' goal is to create a coherent multimodal science fiction story, they should learn how to integrate different modes in meaningful ways, such as adding animations at the end of a textual narrative to convey the message that the story has not ended. Yet, some students were unaware of the integration or did not integrate them efficiently. For instance, students might use visuals and texts to show the same information. Just as it's important to explicitly teach students the connection between disciplinary practices, students are also likely to benefit from explicit instructions on capitalizing on the unique semiotic power of modes and learning different ways to orchestrate modes to integrate different disciplines.

Lastly, educators should select locally relevant science themes and make connections to students' daily lives. Themes with local relevance and connections to students' daily lives can help students think about the themes from a personal perspective. For example, in this project, we chose climate change and human health as the community where students lived needed to prepare for hurricanes yearly. In their stories, we also saw that many groups described situations in which characters responded to extreme weather, such as a hurricane or a tornado. Locally relevant themes are likely to engage students and allow them to connect composing tasks to what they experience in daily life. For example, while we did not explicitly teach climate change indicators, based on their own experiences, students were able to identify these indicators and develop stories about these indicators. We also suggest that educators provide resources for students to learn about the connections between local and global issues. In such a way, students can learn about global issues from a local perspective and learn how to take action to address global issues.

In conclusion, a learner-centered approach incorporating disciplinary role-taking and multimodal composing strategies can help students learn about climate change and develop multimodal science fiction stories with climate change indicators. This approach includes choices in disciplinary roles, modes of expression, and climate change indicators. We encourage educators to provide choices and flexibility for students to use disciplinary practices and modes to develop their expertise and express identities in learning about climate change.

References

Azman, F. N., Zaibon, S. B., & Shiratuddin, N. (2015, November). Digital storytelling tool for education: An analysis of comic authoring environments. In H. B, Zaman, P. Robinson, A. F. Smeaton, T. K. Shih, S. Velastin, A. Jaafar, & N. M. Ali (Eds.), *International visual informatics conference* (pp. 347–355). Springer.

Benwell, B., & Stokoe, E. (2016). Ethnomethodological and conversational analytical approaches to identity. In S. Preece (Ed.), *The Routledge handbook of language and identity* (pp. 66–82). Routledge.

Brown, A. L. (1992). Design experiments: Theoretical and methodological challenges in creating complex interventions in classroom settings. *The Journal of the Learning Sciences, 2*(2), 141–178.

Busch, K. C., Ardoin, N., Gruehn, D., & Stevenson, K. (2019). Exploring a theoretical model of climate change action for youth. *International Journal of Science Education, 41*(17), 2389–2409.

Busch, K. C., & Ayala Chávez, R. (2022). Adolescent framings of climate change, psychological distancing, and implications for climate change concern and behavior. *Climatic Change, 171*(3), 1–19.

Collins, A. (1992). Toward a design science of education. In E. Scanlon & T. O'Shea (Eds.), *New directions in educational technology* (pp. 15–22). Springer.

Eide, E., & Kunelius, R. (2021). Voices of a generation the communicative power of youth activism. *Climatic Change, 169*(1), 1–20.

Jewitt, C. (2008). Multimodality and literacy in school classrooms. *Review of Research in Education, 32*(1), 241–267.

Jiang, S. (2023). Investigating adolescents' participation trajectories in a collaborative multimodal composing learning environment. *Educational Technology & Society, 26*(3), 21–36.

Jiang, S., Shen, J., & Smith, B. E. (2019). Designing discipline-specific roles for interdisciplinary learning: Two comparative cases in an after-school STEM+ L programme. *International Journal of Science Education, 41*(6), 803–826.

Jiang, S., Shen, J., Smith, B. E., & Kibler, K. W. (2020). Science identity development: How multimodal composition mediates student role-taking as scientist in a media-rich learning environment. *Educational Technology Research and Development, 68*(6), 3187–3212.

Jiang, S., Smith, B. E., & Shen, J. (2021). Examining how different modes mediate adolescents' interactions during their collaborative multimodal composing processes. *Interactive Learning Environments, 29*(5), 807–820.

Karsgaard, C., & Davidson, D. (2023). Must we wait for youth to speak out before we listen? International youth perspectives and climate change education. *Educational Review, 75*(1), 74–92.

Kress, G. (2003). *Literacy in the new media age.* Routledge.

Linn, M. C., Clark, D., & Slotta, J. D. (2003). WISE design for knowledge integration. *Science Education, 87*(4), 517–538.

Resnick, M., Maloney, J., Monroy-Hernández, A., Rusk, N., Eastmond, E., Brennan, K., & Kafai, Y. (2009). Scratch: Programming for all. *Communications of the ACM, 52*(11), 60–67.

Sadler, T. D. (2004). Informal reasoning regarding socioscientific issues: A critical review of research. *Journal of Research in Science Teaching: The Official Journal of the National Association for Research in Science Teaching, 41*(5), 513–536.

Sadler, T. D. (2011). Socio-scientific issues-based education: What we know about science education in the context of SSI. In T. D. Sadler (Ed.), *Socio-scientific issues in the classroom* (pp. 355–369). Springer.

Sadler, T. D., & Donnelly, L. A. (2006). Socioscientific argumentation: The effects of content knowledge and morality. *International Journal of Science Education, 28*(12), 1463–1488.

Sadler, T. D., & Zeidler, D. L. (2009). Scientific literacy, PISA, and socioscientific discourse: Assessment for progressive aims of science education. *Journal of Research in Science Teaching: The Official Journal of the National Association for Research in Science Teaching, 46*(8), 909–921.

Sfard, A., & Prusak, A. (2005). Telling identities: In search of an analytic tool for investigating learning as a culturally shaped activity. *Educational Researcher, 34*(4), 14–22.

Shen, J., Smith, B. E., & Jiang, S. (2020). Integrating multimodal composition technology (MCT) in interdisciplinary learning. In L. C. de Oliveira, A. M. Menda, & C. Vincentini (Eds.), *English language teaching methods, approaches, and lessons.* Information Age Publishing.

Smith, B. E. (2017). Composing across modes: A comparative analysis of adolescents' multimodal composing processes. *Learning, Media & Technology, 42*(3), 259–278.

Smith, B. E., Beach, R., & Shen, J. (2021). Fostering student activism about the climate crisis through digital multimodal narratives. *Journal of Sustainability Education, 25.* http://t.ly/2jiF

Smith, B. E., Shen, J., & Jiang, S. (2019). The science of storytelling: Middle schoolers engaging with socioscientific issues through multimodal science fictions. *Voices from the Middle, 26*(4), 50–55.

Zeidler, D. L., Walker, K. A., Ackett, W. A., & Simmons, M. L. (2002). Tangled up in views: Beliefs in the nature of science and responses to socioscientific dilemmas. *Science Education, 86*(3), 343–367.

6

"LISTEN, THERE, TO THE WAY THE REAL WORLD THINKS IN ME"

Cultivating an Empathic Imagination to Support Students' Visual Stories that Address the Climate Crisis

Linda Buturian

Hear Now, a 101-Year-Old Person

I took a break from writing this chapter, and a friend and I went to a Climate Action event *hhh.umn.edu/event/climate-action-together* put on by the Humphrey School of Public Affairs & the Institute on the Environment on the University of Minnesota (UM) campus. It was early August and a beautiful summer day that we Minnesotans cherish; there was music and a food truck, and we had free ice cream from Sonny's and people were making seed paper and screen-printed eco-canvas bags. My friend, Monica *sisterblackpress.com*, pedaled her bike over (Figure 6.1), with the clamshell printing press she attached to the back, and she was helping people print images on postcards with a Wendell Berry quote (Figures 6.2 and 6.3).

Lime sorbet was dripping down my fingers when Dr. Nisha Botchwey, Dean of the Humphrey School of Public Affairs, gathered us together to sing Happy Birthday for someone. About 30 of us clustered around Tom Swain of the Swain Climate Policy Initiative *t.ly/zjPS*, who turned 101 years old. He blew out his candles with a trained effort. Dean Botchwey asked him if he wanted to say anything, and he got up carefully. I just spent a month with my mom, who is 90 years old. 101 is next level. I leaned in. He said something about how each year is a gift and then commended legislators for passing the first climate bill, "The climate crisis is the most important challenge facing humanity today and I urge you all to take action. Thank you."

Minutes before, I didn't know the Dean, Tom Swain, and most people present, so I was surprised at the sudden camaraderie I felt with them, which is an important component of this dynamic which is the climate crisis. Many of us

DOI: 10.4324/9781003335276-9

FIGURE 6.1   Bike Press—Monica Edwards Larson *sisterblackpress.com*

experience the crisis as only out there: Sri Lankans in long lines with no gas for cooking, and Fijians and Kiribatis facing leaving their island homes due to rising sea levels, entire countries enduring famine and drought, glaciers calving, and precious confounding Celsius points. Then the perception shifts, and it is here now, with this group of dedicated people with imagination to dream

FIGURE 6.2   Poetry Spoke Card—Side 1 with Bicycle: Minnesota Center for Book Arts

mnbookarts.org and Monica Edwards Larson

FIGURE 6.3   Poetry Spoke Card—Side 2 with Tree and Quote: Minnesota Center for Book Arts

mnbookarts.org and Monica Edwards Larson

up the event, fund it, and gather diverse artists and activists. In the presence of a centenarian whose life spans two pandemics, a world war, and the Great Depression, using his precious breath to direct our focus on the climate crisis. This movement from out there and overwhelming to here now, even inside of us, and a feeling of hope, takes knowledge, vision, imagination, creativity, playfulness, empathy, and labor. I will write about these essential components to my teaching about the climate crisis and the ongoing struggle for a sustainable planet.

What follows is a narrative of how I designed assignments in two undergraduate courses, an arts education course and a Global Literature course, that meet liberal education requirements at the University of Minnesota and have nothing seemingly to do with climate. You will learn about how my curricular integration is part of an initiative by my Department, Curriculum & Instruction, in taking the lead on The Center for Climate Literacy *climateliteracy.umn. edu*.

You will become acquainted with my students, Gisell, Lillian, Henry, and Chun, to see how, in Tom Swain's words, we take action on the climate crisis. I caught up with the four of them via email and asked each of them about their engagement with the assignment within the context of the course. I also asked them what impact, if any, this has had on their academic learning, as well as what they would like the University and you, the readers, to know. I am grateful to them for agreeing to share their creative work and writing, as well as their thoughtful answers. My aim is to provide you with my (our) story to further equip fellow educators. I welcome your stories, assignments, and responses.

Art for the Climate: Art for Social Change Capstone Assignment, CI1032 "Creating Identities: Learning In and Through the Arts"

Gisell

We were a few weeks into the semester before I realized that Gisell was a PSEO student; she was in high school in Minneapolis and taking my course for college credits. For Gisell's Art for Social Change assignment, she chose to create a mural depicting the intersection of climate justice and her Chicanx culture (Figure 6.4). She drew culturally potent images and selected vibrant colors reflecting her Chicanx heritage and the fight for environmental justice. Her distinct pen marks reveal the effort involved in the struggle and how we make progress stroke by stroke.

In drafts of her artist statement *t.ly/PKM4*, Gisell had no problem describing her mural and communicating what the title, symbols, and colors signify. What

FIGURE 6.4 *"La Lucha Sigue. ¡Si Se Puede!"* by Gisell Ayala-Corral

she found challenging was how to define climate justice and connect it to her Chicanx culture, partly because she was absorbing the intersectional nature of climate justice.

In her words, "Growing up, I was never really interested in the issue of Climate Change. I thought that people who were passionate about it were outdoorsy and often went camping or were just constantly doing stuff outside. I was none of that growing up. I live in North Minneapolis. A place where there is little to no space for you to be outdoors and enjoy nature."

Gisell's struggle is common among the students who choose to tackle climate change for their capstone in my course, CI1032 "Creating Identities: Learning in and Through the Arts."

I share their struggle; my background is in the humanities, in literature, and in creative writing. I am an environmental activist and have designed and taught courses on oil dependency and its global impact, and on water *cehd.umn.*

edu/PSTL/water, where each student created digital stories addressing a water resource topic, as well as integrated environmental themes in my teaching *openrivers.lib.umn.edu/article/the-river-is-the-classroom* (Buturian, 2018). However, climate change is not my research focus, and the urgency, scope, complexity, and pace of the crisis overwhelm me. I share this struggle with other educators who are not climate/environmental scientists and who are integrating climate literacy curricula into our courses.

Gisell and I and tens of thousands of others are on the University of Minnesota, Twin Cities campus built within the traditional homelands of the Dakota people. Like many other educators, I begin my classes (and this chapter), acknowledging that this land was taken by force, and I continue to find ways to make this unresolved truth have heft and substance. I will return to this later in this chapter as it relates to the climate crisis.

Gisell was one of 30 students in "Creating Identities: Learning In and Through the Arts," an analytical and experiential course that explores how art participates in our understanding of the intersection of race, class, culture, ethnicity, religion, gender, and the environment. No prior artistic experience is required, and all students are welcome. Because the course satisfies a liberal education requirement as well as an elementary education major elective, the class has a mix of teacher candidates and students from throughout the UM, from undeclared first-year students to seniors in diverse majors. This translates into a more culturally and ethnically diverse class, though still predominantly white.

I have designed the major assignments so that students have agency over what they choose to focus on and how they will create. I've been teaching some iteration of this course for a decade and have integrated movement and dance and an emphasis on social justice. My pedagogy is shaped by an assets approach and youth participatory action research. Thus, assignments invite students to consider their lived experiences and identities and to activate their creativity and involvement in social action. Students pair images and narratives to build on their visual literacy and further their ability to think and communicate metaphorically.

I designed the culminating assignment, Art For Social Change, with graduate student Jessica Hron. The assignment asks students to choose a social justice topic, conduct research and fieldwork, and select a creative medium they want to work in. They then bring the topic and medium together to create art that educates and inspires, as well as craft an artist statement that includes relevant information, both about the topic and the artist and posits a strategy for addressing the problem. (See article *t.ly/KUzt* for more examples.)

Students brainstorm topics and complete assignments to help narrow the topic and stay on track—discreet due dates and written assignments for research, fieldwork, and statement drafts. Students choose topics including race and gender stereotyping in Division 1 athletics, the school-to-prison pipeline, the Twin Cities as a hub for human trafficking, preventing sexual assault on campus, understanding the needs of unhoused youth, food scarcity, both on campus and globally, transgender peoples' lack of access to health care, and more. Creative mediums include choreographing and performing a dance, spoken word poetry, textile art, painting, photography, digital stories, and music. Some collaborate, and others work solo. Finally, students premiere their art and statements for their class.

Every semester, a few students focus on environmental justice and climate change. Their art educates their classmates and becomes a model for future students. Back in 2016, the first student to tackle climate change choreographed a dance, which she performed in the woods in winter along the Mississippi River. You can see segments of her video here *t.ly/8Tpi* (Buturian, 2019).

Back to Gisell, who was my student in 2019, and who graduated in 2022 with a BA in mathematics. She will continue her graduate studies at the University of Minnesota and pursue a PhD in mathematics starting in the Fall of 2023. Here is Gisell's response to the question of the impact of creating art about the climate crisis and what other courses she took that addressed climate change:

> Over the course of my academic career, I have taken some classes that touch on the subject of climate change/global warming, but I don't think they really went into depth, it was more of the instructor acknowledging that we have a problem with our climate... I think CI 1032, the class I took with Linda, was one of the few classes where it allowed me to learn more about it and helped me express how I feel about this big issue that our world is facing. I thought that it was very interesting that this class allowed us to explore whatever issue we wanted to (I chose climate change) and use art as a means to convey a message. We did a bit of research and used our own lived experience to produce an art piece we'd present in the end.

> My biggest takeaway from this assignment was that art can have a huge impact on people. I've realized that art captivates the attention of people like nothing else. It also makes them more engaged in learning about the problems artists may be talking about through their piece... I remember this being a very enjoyable activity because as I did my piece, I was reflecting a lot on my identity and my relation as a Chicanx woman and how my community is affected by climate change. Art is a very powerful tool because it allows for creativity to flow and for reflection to occur.

Gisell's insights on the joy she felt reflecting on her cultural identity is an essential component in our understanding of the climate crisis, which I'll return to later in this chapter.

In response to what impact this engagement with climate change continued to have, Gisell wrote:

Since I have been very passionate about climate change, and since I didn't get a lot of exposure to learn about this problem, I took it upon myself and learned about it through research. Much of the research that I've done has been through a mathematics lens. I've been looking at the data from the ocean heat content in relation to atmospheric CO2 and trying to come up with a mathematical model that reflects what is happening between the two since these two factors greatly affect our climate system. This may not necessarily reflect the social aspect of the issue, but it is hard data that indicates that there is a big problem that our planet is facing. I would like to use what I learn from my research to generate solutions to this problem. I would like to add my insight as a Chicanx woman since often my experiences aren't reflected on the table when it comes to climate change due to a lack of representation.

Gisell's insights help me consider the arc of impact, the interaction of my course with other courses, and the need for more educators to integrate climate issues in their respective disciplines.

Lillian

When I found out that Lillian was an Environmental Science major, I jumped up and down. Her knowledge grounded our class's discourse around climate issues. Lillian was a student in my spring 2022 class and is now a Senior majoring in Environmental Sciences, Policy, and Management, with a focus on Policy, Planning, and Law. Her minor is in Sustainability Studies, which means almost every class she has taken has focused on climate change. As a result, her classes approach climate change from differing lens, including environmental justice, ecology, policy, economics, and systems thinking.

However, Lillian had never been asked to or attempted to create art in response to climate issues. For her Art For Social Change project, she focused on environmental injustice and the enduring impacts of redlining, which she defines in her Artist Statement *t.ly/γ2v-*.

Lillian's creative medium evolved as she considered the best way to communicate her research findings visually. She was also keen to try her hand at welding. So, she chose a three-dimensional bird's-eye view (Figures 6.5–6.7).

Lillian's response to my question about the impact of her experience illuminates the unique role that creative expression plays. Her words follow:

This project allowed me, and really encouraged me, to incorporate my feelings about the climate crisis into the work in an unrestricted way. It

FIGURE 6.5   *Invisible Lines* by Lillian Prybil

FIGURE 6.6   *Invisible Lines* Close-up with Lake by Lillian Prybil

FIGURE 6.7 *Invisible Lines* Close-up 2 with Fist by Lillian Prybil

challenged me to go out of the comfort zone that I got used to after over a decade of being told that learning at school isn't about emotions. I initially struggled with how to present my feelings toward the topic of environmental justice in a visual manner.

Though I knew how to express my emotions about redlining and its aftermath in words, I didn't really know how to show the weight of this without words. It felt difficult to express themes such as racism and resilience from a bird's-eye view of a city. I learned firsthand the importance of art to communicate complicated ideas like climate justice. Not only did I learn how to use art to communicate with others, but I also improved at bridging the gap between my more "objective" or "academic" thought processes about a topic and my creative thought process.

Lillian's phrase, "bridging the gap" between analytical and creative thinking, brings to mind Juanita Schlapfer-Miller's (2017) writing on how humans process information about climate change in her chapter, "Climate Garden 2085: An Art Science Experiment Promoting Different Ways of Knowing about Climate Change" (p. 152).

"However, in analyzing human responses to climate change, psychologists have shown that we use two different processing systems. The first is intuitive, experiential, affective (emotional), and fast. The second is deliberative, analytical, rational, and slow. We constantly make judgments using these systems in parallel, but when they diverge, the first system dominates. In other words, how we feel about something has a stronger influence on how we respond (Slovic & Peters, 2006)" (Schlapfer-Miller, 2017, p. 152).

What Gisell and Lillian are enacting in their research, combining it with their lived experience and then creating art and writing, activates both "processing systems" (Schlapfer-Miller, 2017, p. 152).

In my book, *The Changing Story: digital stories that participate in transforming teaching & learning* (Buturian, 2016), *https://pressbooks.umn.edu/thechangingstory*, I explore this important bridge-gapping when students create digital stories, which holds for art about climate change. I reference Bloom's Revised taxonomy *t.ly/rTch* among others, where "creativity" is deemed a "higher order thinking skill" at the pinnacle of the learning triangle. Colleague Scott Spicer's "Hierarchy of student-created media" helps us visualize the learning process these assignments foster in students (see Oziewicz and Spicer's Chapter 1 for the figure).

When students move through these kinds of learning about, in this case, the climate crisis and then create visual and written work that communicates their findings to others, they engage in potentially transformative learning that, as Gisell's and Lillian's responses reveal, can lead to further action and engagement.

It's modest, in the broad scheme of things: a few students focusing on the climate crisis out of a whole class. However, Gisell and Lillian's reflections help us understand the potential enduring impact as well the effect on their classmates. This cross-pollination of students' interests and loves amplifies and educates. Our Department offers five sections of the course every semester, meaning hundreds of students engage with climate literacy and other social justice topics.

In some cases, this might be the only encounter students will have, as we'll soon learn from Henry, in the "Global Stories of Education: Literature for Young Adults" class. And the assignments I'm integrating contribute to a larger climate initiative our Department is leading, which I'll also address in the next section, and as described in Marek Oziewicz and Scott Spicer's Chapter 1. This also points to the need for other educators to integrate climate curricula into their respective disciplinary teaching.

The Village That Is My Department: Curriculum and Instruction

The complex, intersectional nature of the climate crisis requires us to collaborate. My Department's commitment to environmental justice rises out of our ongoing race and equity justice work, which is central to our mission. Through work groups, readings, and teach-ins with Bipoc scholars, our predominantly white faculty have been unpacking the impact that the legacy of forced slavery of Africans, and the removal and genocide of Indigenous peoples have on our society, selves, teaching, and interactions with students and others.

The murder of George Floyd by police, which happened a few miles from our campus, further catalyzed colleagues, staff, students, and myself to unpack how white supremacy is perpetuated in our teaching, research, and engagement with each other and students, as well as how whiteness, colonialism, and imperialism inform our perceptions and actions. We are centering the voices and knowledge of scholars, educators, and activists of color to cultivate a truly welcoming and sustaining community for Bipoc colleagues, staff, and students. We need their knowledge, experience, and vision to move forward on equity in race and the environment.

This work of centering race and justice has led to an intersectional approach to understanding and teaching about environmental justice and climate activism. Within the Department, there is a group of colleagues and grad students who are integrating climate justice into their respective courses and research. Marek Oziewicz, with the support of the Chair, Dean, and Department, launched the Center for Climate Literacy *climateliteracy.umn.edu*.

Environmental justice is a part of our mission as well as how it was recently included in a faculty description search. As the education department, we teach the elementary education major and have relationships with K–12 public schools across the state and in the Twin Cities where we place our interning students; we share curricula, which impacts future teachers. (See, for example, Climate Lit *climatelit.org* as "a hub for K-12 resources for educators to advance climate literacy.")

Because we are still a predominantly white faculty, we must center Bipoc scholars, writers, artists, climate scientists, farmers, and activists in our teaching and engagement. A few to note that I am learning from and integrating into my teaching are Dr. Ayana Elizabeth Johnson *ayanaelizabeth.com*, Dr. Chandler Puritty *t.ly/MezU*, writer/activists adrienne maree brown *adriennemareebrown.net*, Erin Sharkey *t.ly/Pve3*, the All We Can Save project *allwecansave.earth*, The Joy Report *intersectionalenvironmentalist.com/the-joy-report*, and the Afro-Indigenous-centered community, Soul Fire Farm *soulfirefarm.org*.

Race and equity work includes supporting, amplifying, and learning from the work of Indigenous people. Led by Ojibwe colleague Dr. Mary Fong Hermes, our Department is welcoming five Indigenous scholars in Fall 2022, who will be completing PhDs in Education, in classes that will be taught in both Ojibwe and English.

My hope is to connect my learning from these scholars with students in the two classes I've written about here. If the scholars want to meet with my students, they can hear directly from Indigenous women about their understanding of the climate crisis. In the Global Literature class, I was fortunate to take part in bringing Chickasaw writer Linda Hogan into the Global Literature classroom (virtually, due to the pandemic). Students read essays from her collection, *The Radiant Lives of Animals* (Hogan, 2020), and selected poems, and were moved and inspired by the stories of her deep understanding of the land and animals she lives with and among.

Early on in this chapter, I acknowledged that the University of Minnesota's Twin Cities campus is built on Dakota homeland. I share this passage by Čhaŋtémaza (Neil McKay) and Monica Siems McKay (2020) from the University of Minnesota's *Open Rivers* journal, *openrivers.lib.umn.edu/article/where-we-stand* that illumines how the Dakota's absence and presence connect to the climate crisis:

> The Dakhóta connection to the land and all that live and exist here is important. The Dakhóta people and other Indigenous peoples have seen for thousands of years that we must be aware that we co-exist with other life. Human beings are not the most important life on earth; in fact, we can't survive without help from our relatives, but they can manage quite well without us. The Dakhóta philosophy of Mitákuye Owás'iŋ, "all my relations," or "I am related to all that is," reflects this understanding by acknowledging that all things from water, plants, and animals to the stars are part of our fellow creation and we must maintain a respectful relationship with all of these things we are connected to. This brings us back to the observation that traditionally, the Dakhóta and other Indigenous peoples did not construe their relationship to land in terms of ownership, but rather of belonging and stewardship. Again, we mention this not to romanticize Indigenous people, but rather to suggest that we can peel back the layers of legal sleight-of-hand through which, as Martin Case (2018) puts it, Indigenous land was transformed into U.S. property; if we can return treaty lands to their rightful owners; then we open up the possibility of paying the lands and waters themselves, as well as the lands' original inhabitants, the respect they are due.
>
> *p. 19*

Stories as Homage to the Earth: The Homage Assignment, CI 1124 "Global Stories of Education: Literature for Young Adults"

"Global Stories of Education: Literature for Young Adults" is a course designed for students in majors outside the humanities and enrolls many upper-level students in the sciences, engineering, nursing, and business, as well as first-year students and sophomores in diverse majors. It is a delight to teach as students tend to have strong academic skills; yet, they have not had much chance to read novels, engage their creativity, and contemplate the vital role that culture plays in both engaging with and understanding literature (and climate change).

In addition to teaching a section every semester, I supervise graduate students who also teach this class, and our collaboration is mutually enriching in furthering our understanding of education. With Marek Oziewicz, we are integrating a climate literacy theme in all sections of CI1124 and designing new literacy courses focused on climate change as part of the work of the Center for Climate Literacy *climateliteracy.umn.edu*.

In the literature course, this translates into selecting literary texts by culturally and ethnically diverse writers from throughout the globe. Students complete assignments that address humans' interaction with the natural world; how colonialism, capitalism, and imperialism shape our cultures and imaginations; how our dependency on fossil fuels drives the crisis; and the impact on us, ecosystems, and biomes.

The final assignment, *Homage*, asks students to produce a creative response in homage to a theme or specific literary text. They choose what they want to focus on and what creative medium they would like to work with, and then write a reflection that helps shed light on the connections between their creation and the theme/text. Students have chosen music composition, spoken word, short story, three-dimensional art, poetry, baking and cooking, textile art, multimedia, woodworking, and painting.

Henry

Henry was in his final semester as a student in my Global Literature class last Spring 2022. In May 2022, he graduated with a degree in Interior Design. He had to rush from St. Paul to the Minneapolis campus to get to class and would often slide into a front seat just as we were starting. Like most students, Henry took the class to fulfill the liberal ed requirement.

Henry enjoyed reading the novels and poems and was especially taken with Becky Chambers' solar punk novel *Psalm for the Wild-Built* (2021), a futuristic novel where civilization collapsed, and those who survived established sustainability for all. In an interview with the author that we watched in class, Chambers shared that *Psalm for the Wild-Built* was influenced by biologist E. O. Wilson's book *Half Earth* (2016). Wilson proposes that half of the Earth's surface should be designated a human-free reserve for all other species to sustain biodiversity.

Inspired by this premise and using online tools, Henry designed Panga Cafe *t.ly/hXEMn* which includes a QR code *t.ly/B7wx* so that you can view the cafe virtually (Figures 6.8–6.11). Here is his written reflection *t.ly/5x24*.

FIGURE 6.8 "Panga Cafe" QR Code by Henry Nguyen

api2.enscape3d.com/v3/view/f9697ae2-3996-498c-b9ef-73bba73d96e6

FIGURE 6.9 "Panga café" Front View by Henry Nguyen

In response to my question about what other classes in his academic career addressed climate change, Henry wrote: "Only 1 class and that is Linda's class that addressed this issue." We need more educators and more integration of climate curricula.

Chun

Chun, our final student, is currently a junior in Computer Science. In my conversations with him before class started, he shared that he is from Washington and spends portions of his summers hiking different sections of the Pacific Crest Trail (PCT). I told him how I used to live along the Trail in the mountains

FIGURE 6.10 "Panga Cafe" Rear View by Henry Nguyen

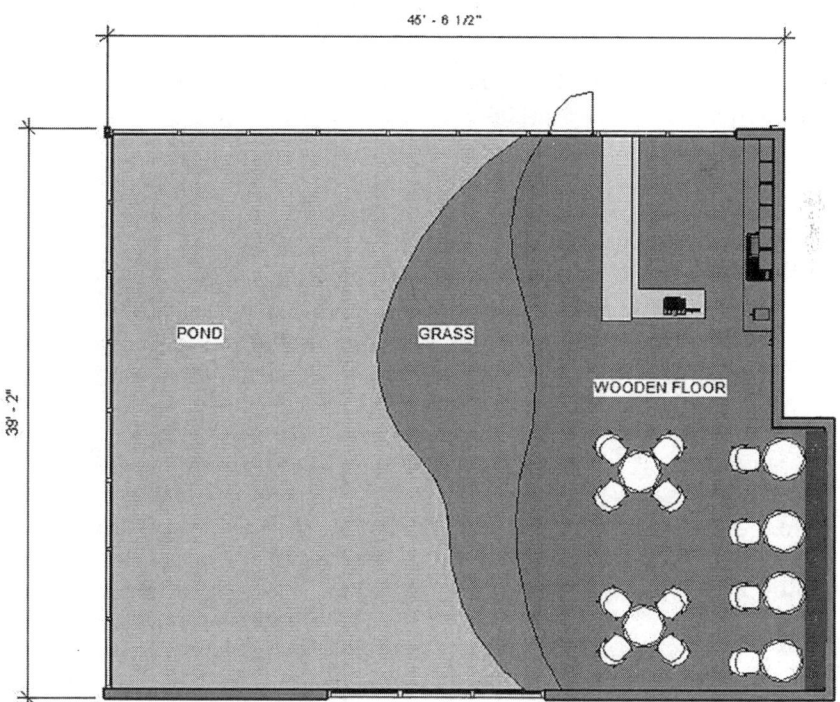

FIGURE 6.11 "Panga Cafe" Floor Plan by Henry Nguyen

of Southern Oregon and that, in my environmental work, we succeeded in keeping logging from happening along that portion of the PCT. In his written reflection *t.ly/ujal*, Chun resonated with the environmental/climate theme in our readings and discussions, including an Afro futurism module with an Octavia Butler short story, music video, podcast, and art, as well as references to *The Little Prince* by Antoine De Saint Exupery (2000), *Psalm for the Wild Built* by Becky Chambers (2021), and *Me, Who Dove Into the Heart of the World* by Sabina Berman (2012).

Chun was the first student in the Global Stories of Education who took me up on the option to write a short story for the Homage assignment, and it was his first time writing a story. I have included the opening two pages here. Follow the link to read his full short story *t.ly/j8Ha*.

It Was (Is) Beautiful
By Chun Lin

There are some things that my grandma used to always say to me.

"Live like the land around us."

"We were born to have our own purposes, callings that tugged at our hearts, begging to be answered someday, so it would be a shame to not do so. Even at the small place we call home that no one else could see, on the peak of the tallest mountain, if we lived like the land, wouldn't we be happy our entire lives?"

"If we lived like the land, wouldn't our purposes become even more radiant?"

I used to always smile confusedly and nod, having heard this countless times, not knowing what she meant by it all. In the back of my mind, wasn't staying just like this, the two of us, perfect? Why would we ever need to do anything else to be happy?

Some Mountain Spirits lived longer than others; depending on their soul, they could maybe live even as long as the great pine trees, or they might shine their brightest over the course of a week before burning out. They stay as long or as short as they do because every Mountain Spirit is born with a purpose that manifests itself sometime during their life as a journey their soul must embark on. When a Mountain Spirit fulfills that purpose and reaches the end of their journey, they dissipate. They disappear without a trace, like the words spoken were never heard in this world and the only things left of their existence would be footprints and belongings that would only get lost in time.

Chun has taken several science/environmental classes that address climate change "extensively" on environmental ethics, justice, and policy. He also took a lab on renewable and sustainable energy with an overarching focus on climate change, including an energy audit where he had to measure his energy use. When I asked him if and how writing a short story in response to the climate theme differed/contributed, he wrote:

It helped me personalize the ideas surrounding climate change. Taking a creative response, showing what inspires us, and creating something more abstract, is really different from a direct lesson or reading on climate change facts and then taking a quiz on its contents.

Culture and the Climate Crisis

Our perceptions of climate change and the crisis are culturally bound. This notion that the crisis is "out there," which many of us in the Western world feel (unless you live on the West coast and are facing fires or along the ocean shore watching your coastline disappear), is a cultural construct. It is hard to help students understand that their perceptions are culturally inscribed, like trying to catch a glimpse of their own shadows. For example, in Minnesota, the land of more than 10,000 lakes, part of our cultural norm is that there is plenty of sparkling clean water. And yet, in 2019, 56% of those lakes were classified "impaired."

In 2022, the Minnesota Pollution Control Agency (2022) added 305 water bodies with 417 new impairments, bringing the total to 2,904 water bodies with 6,168 impairments (MPCA). From the number of days earlier the ice is going out, to the frequency of algae blooms and invasive species in our water bodies, climate change is here now, impacting our way of life, and yet many have an outdated understanding of water. Further, our Western concept of nature being out there, separate from us, perpetuates this distancing, evidenced in phrases like "save the planet" as if it is an entity apart from us.

In the Global Literature class, we spend time discussing the role of culture in the stories and poems we read and our lives. (I designed an assignment, "Culture in a Story," where students share a cultural tradition they are a part of. Many white students struggle with feeling that they even "have culture" as they say. They think that culture is what people with brown skin have, and their culture is neutral, or "just like everyone else's.")

Engaging with stories can help us understand the cultural confines of our understanding of nature and climate. My chapter title comes from Karen Nieto, the main character and narrator of Sabina Berman's (2012) novel, *Me, Who Dove in the Heart of the World* (translated from Spanish by Lisa Dillman). Karen Nieto creates an instruction manual for herself in order to function as a person with autism while also helping her aunt run a tuna cannery in

Mazatlan. Karen's instruction to herself is, "Listen, there, to the way the real world thinks in Me" (Nieto refers to herself as Me throughout the novel).

Nieto takes issue with the damaging impact that Cartesian thinking has had on our imaginations by prioritizing thinking before and over existing. She states that due to this prioritizing: "Trees, the sea, the fish in the sea, the sun, the moon, a hill or a whole mountain range. None of that exists all the way; it exists on a second plane of existence, a lesser existence. Therefore, it deserves to be merchandise or food or background for humans and nothing more" (p. 109).

Karen's blunt refusal of Descartes's philosophy, combined with her love of swimming with and understanding tuna, and where that affinity leads her, draws students into her world. Her character raises questions that foster conversations about our relationship to the non-built environment, and how our separation from nature, due to technology, capitalism, colonialism, and continued dependence on fossil fuels, participates in the slow (and fast) violence. Karen, in addition to Greta Thunberg, Temple Grandin, and others, also helps us consider the role that neurodivergent people can play in our understanding of nature and climate.

The Role of Empathy in the Fight for a Sustainable Planet

Much has been written about how engaging with literature can foster empathy for others. For example, I have students read philosopher Martha Nussbaum's chapter on "Democratic Citizenship and the Narrative Imagination" *t.ly/BPWp* (Nussbaum, 2008). She develops how reading classics helps us access others who differ from our way of thinking and being and how this reading is linked to active citizenry. It is profound the impact engaged reading has on our students.

In our Global Literature class, the power of the writing combined with Karen's humor, honesty, and resilience connects students to Karen, and through Karen, the ocean and tuna, in a palpable way. A student commented after reading *Me, Who Dove* that she gave up eating tuna, and is considering vegetarianism. This points to what I love about engaging with young people; that they are open and want to learn, to be challenged, and to change according to what they discover.

More of them are coming in with experience in social action and media creation from high school. Are we ready to support them so they can see how their academic skills participate in the struggle?

Empathy, then, involves understanding and honoring our interconnectedness, the Dakhota philosophy, "all my relations" (McKay & McKay, 2020, p. 19). Students need to have an empathic imagination for the climate crisis, for humans, animals, and ecosystems.

Storytelling's Role in the Climate Crisis

Stories are at the center of both courses I have shared and storytelling is an essential element in the fight for a sustainable climate and planet. In Kim Stanley Robinson's (2020) relevant work of fiction, *The Ministry for the Future*, he writes: "A just civilization of eight billion, in balance with the biosphere's production of the things we need; how would that look?" (p. 320). To envision how this would look, we need to foster radical imaginations, creativity, collaborative learning, and hope. Students also need to learn the science, economics, policy, and politics relevant to understanding global warming. Students must be able to craft stories in diverse mediums that share their passion, concern, knowledge, vision, and strategies for creating a sustainable future.

Implications for Teaching

If readers are new to integrating climate literacy into their teaching, consider starting modestly. For example, design a low-stakes assignment such as students choosing a relevant image from the media or one they create, which they analyze and reflect on. It took years of experimenting and teaching for me to gain comfort in allowing students to select creative mediums. Begin with the mediums you are comfortable with. When you design a new assignment or expand the options students can choose from, share the risk with them. You could make the assignment worth points that they receive if they complete the assignment. Once you have some models, and more clarity about what it is that you want from them, and how best to communicate that to your students, you can then start fine-tuning the assessment. Consider involving your students in creating the assignment and in building the assessment:

- Collaborate. Ask for help.
- Model strategies for addressing the crisis in order to sustain your hope as well as theirs. Require solutions in addition to understanding the problems.
- Students are eager to work with their hands and do DIY. Create experiential assignments such as gardening and designing alternative power systems or collaborate with those implementing these projects.
- Find ways to connect students with people across the globe who are directly impacted by the crisis.
- Discover good reasons to go outside.

"Listen, There, To the Way the Real World Thinks in" Them

I conclude this chapter with responses from Gisell, Lillian, Chun, and Henry to my question: "Most of the readers for this book are educators. What would you like them and/or the UM to know?" Thank you for listening to them.

Gisell

I think something that I would like to convey to educators that may be reading this book is for them to talk about their lived experiences when teaching. I think when they talk about their experiences it makes the content of what they are teaching more relatable and it makes it more engaging, and this becomes very helpful, especially when talking about issues like climate change.

Lillian

I wish we were asked to simply sit in on the local governance process more often. While we were often called to make plans for improving governance or propose engagement processes for making climate policy decisions that benefit the residents of the city/county, I didn't truly understand what the existing local governance process was like. I learned a lot more about the political process from watching a city council committee meeting during the first week of my internship than I did from reading about policy change and watching lectures in class. Taking this further, I would have enjoyed communicating with my elected officials through visual art, writing, or speaking. Instead of simply being lectured at and told to engage in the political process, it might have been nice to have an assignment that required you to become more politically engaged.

Chun

One of the key takeaways from my understanding of climate change is that as individuals in society, we consume way too much. Inside the classroom and in our community, I'd like to see more active choices being made for greener alternatives. Whether this be biking to class instead of driving, using metal straws and reusable bottles, switching from fluorescent to LED bulbs, and more.

Henry

More awareness of climate change and how to protect our planet. We were taught to exceed in our own path and our own major, but we were not taught how to treat other people well or how to care for our planet. That is what I would love to see more in the school setting in the near future. I have seen many people with a four-year degree or a master's degree who do not know how to act as a good citizen, or as a well-behaved human being. They can still throw trash on the street or treat other people, the environment, and other animals badly. I think it would be beneficial to integrate some ethical lessons into the lectures to give more awareness to the students. I think educated people have to be the ones that not only have good grades but also good souls.

References

Berman, S. (2012). *Me, who dove in the heart of the world* (trans. L. Dillman). Henry Holt & Co.

Buturian, L. (2016). *The changing story: Digital stories that participate in transforming teaching & learning.* College of Education and Human Development, University of Minnesota. https://pressbooks.umn.edu/thechangingstory

Buturian, L. (2018). The river is the classroom. *Open Rivers: Rethinking Water, Place & Community Journal, 10.* https://openrivers.lib.umn.edu/article/the-river-is-the-classroom/

Buturian, L. (2019, February 1). Visual narratives about environmental justice. Presentation at the Sustainability Education Summit, Institute on the Environment, University of Minnesota. https://www.youtube.com/watch?v=sCBgnwH6HXw&t=1599s

Case, M. (2018). *The relentless business of treaties: How indigenous land became U.S. property.* Minnesota Historical Society Press.

Chambers, B. (2021). *Psalm for the wild built.* Tordotcom.

De Saint Exupery, A. (2000). *The little prince* (trans. R. Howard). Clarion Books.

Hogan, L. (2020). *The radiant lives of animals.* Beacon Press.

McKay, N., & McKay, M. S. (2020). Where we stand: The University of Minnesota and Dakhóta treaty lands. *Open Rivers: Rethinking Water, Place & Community, 17.* https://openrivers.lib.umn.edu/where-we-stand

Minnesota Pollution Control Agency. (2022, March 8). MPCA Waterfront Bulletin. http://www.t.ly/_Bql

Nussbaum, M. C. (2008). Democratic citizenship and the narrative imagination. In G. D. Fenstermacher (Ed.), *Yearbook of the national society for the study of education, 107*(1), pp. 143–157. Wiley Press. doi:10.1111/j.1744-7984.2008.00138.x

Robinson, K. S. (2020). *The ministry for the future.* Orbit Hachette Book Group.

Schlapfer-Miller, J. (2017). Climate Garden 2085: An art science experiment promoting different ways of knowing about climate change. In M. Dahinden & J. Schlapfer-Miller (Eds.), *Climate Garden 2085: Handbook for a public experiment* (pp. 145–165). University of Chicago Press.

Slovic, P., & Peters, E. (2006). Risk perception and affect. *Current Directions in Psychological Science, 15*, 322–325.

7

ADDRESSING CLIMATE CHANGE AND SUSTAINABLE ENERGY FUTURES THROUGH CREATIVE MUSIC ENGAGEMENT

Evan S. Tobias, Kyle Bartlett, Michelle E. Jordan, and Steven J. Zuiker

Introduction

The severity of the climate crisis cannot be overstated. According to the Intergovernmental Panel on Climate Change (IPCC, 2022), we have until 2030 to avoid or mitigate the worst impacts of climate change. The urgency and complexity of the climate crisis require us to expand from addressing it to modifying the type of thinking that led to climate change. Education is a powerful agent for increasing students' environmental literacy, interdisciplinary collaboration skills, multimedia communication practices, and awareness of systemic interdependencies to support such change (Beach et al., 2017; Littrell et al., 2020). Young people worldwide are leading climate change and ecosystem sustainability activism, challenging current social, technical, and energy systems that foster unsustainable practices and overly burdening poor and marginalized people (Drehobl & Ross, 2016; Zraick, 2019). Young people are also exhibiting sophisticated future perspectives that stretch beyond imagining their future selves, to also imagining dystopian and hoped-for futures for themselves, their communities, and world ecologies that transcend their lifespans (Jordan et al., 2021).

Educators should consider centering and amplifying youth perspectives about their futures and the futures of their communities. This means making space in curricula and classrooms for students to engage with pressing issues around climate change and supporting students' design of more sustainable and equitable futures. As Fisher and Mahajan (2010) argue, education should support students in inventing and designing the future and foster capabilities ranging from framing problems to working through uncertainty needed to transform and thrive in such a world.

DOI: 10.4324/9781003335276-10

This chapter highlights how educators can inspire and enable students to shape their futures by combining Education for Sustainable Development (ESD), futures thinking, and music creation. Involving young people in creating possible futures for their communities through music engagement can increase youth interest in and identification with science and art. Doing so can also shape the future by bringing together youths' shared commitment to clean energy activism through a common focus on energy systems transitions.

Wicked problems such as addressing climate change require interdisciplinary approaches (Giangrande et al., 2019) and call for students learning, creating, and collaborating in ways that draw upon diverse expertise. The approach that we describe couples science with arts (music specifically), where both disciplines build upon one another as students work through real-world problems around climate change. This differs from an approach in which educators add music as a separate or subservient component to science projects, such as limiting music to the role of accompanying facts about solar energy (Bresler, 1995). Here, music and science are equal partners in an integrative approach (Kysilka, 1998).

Educators of all kinds may be interested in supporting such transformational learning agendas. They might consider questions such as:

- How might combining ESD, futures thinking, and music engagement play a role in addressing climate change?
- How might students engage with music to imagine future energy transitions to support decarbonization in their communities?
- What roles might educators play in supporting students to imagine possible futures that are sustainable through music?
- What might youth have to communicate about sustainable futures through their musical creations and the activity of creating?

Drawing on our research that involved high school students creating possible solar futures through music and instrument creation, we describe one way educators might integrate ESD practices, futures thinking, and music engagement in schools. While the project we highlight focuses on music engagement as a medium for futures thinking around energy transitions, we invite all educators interested in addressing climate change through music to consider related interdisciplinary and collaborative possibilities that could be meaningful in their programs.

Climate Change Impacts and the Role of Education

Ensuring that all people, especially those who are minoritized or who identify as women and girls, receive a quality education is crucial for global climate stabilization efforts and sustainability. Education increases local and global environmental awareness, helping learners understand current environmental concerns ranging from deforestation to droughts, and empowers learners with knowledge of local policies and how to engage in policy transformation such as voting and activism (UNESCO, 2014). In addition, arts education can provide youth with a creative and expressive medium to work through these issues. Paired with ESD *unesco.org/en/education-sustainable-development t.ly/7ews*, a framework to help integrate sustainability issues into all subject areas, an integrated approach to arts and science education offers a powerful means for preparing youth to address increasingly complex environmental, economic, and social challenges.

Education for Sustainable Development

ESD is a framework developed by UNESCO (2022) for preparing students with the skills, values, behaviors, and agency required to address local and global challenges and ensure that future generations can flourish. ESD emphasizes student agency and social transformation locally and globally by empowering students to make decisions, care for the planet, and act for the betterment of social, environmental, and economic systems. Educators seeking to address climate change in their programs might apply ESD to their existing curriculum by having students (1) work with their subject area through social, environmental, or economic lenses and (2) address global and local challenges.

This chapter's particular content case in point involves K-12 music programs. For instance, students in ensembles and music classes could research the raw materials, labor, transportation costs, and impact on the planet involved in the manufacturing and delivering instruments and equipment with which they make music. Students might then discuss the implications of and responses to what they learn. This approach addresses concepts in the existing music curriculum while supporting students to develop systems thinking capacities, a key component of ESD, and developing a deeper awareness of the relationship between their lives and issues of sustainability.

The urgency of climate change calls for educators to expand from focusing on awareness of or responses to climate change through music to fostering students' ability to transform the world in ways that protect the planet and enable all its inhabitants to flourish (United Nations Educational Scientific and Cultural Organizations & Mahatma Gandhi Institute of Education for Peace and Sustainable Development, 2017). Empowering youth to transform the world calls for educators to support students in imagining just, equitable, and sustainable futures. Futures thinking, a key aspect of ESD, is a way for students to consider possible futures of the world and imagine the worlds they wish to inhabit and that their communities need to thrive. The project we outline in this chapter couples science and music as a means for students to practice futures thinking to imagine sustainable worlds powered by sustainable and equitable energy systems as a step toward positive transformation.

Futures Thinking

Operating from the belief that the future emerges from the world and ourselves in interaction, futures thinking is an interdisciplinary collection of theories and approaches to help people engage in "the rigorous art of imagining" (Miller, 2003). Engaging in futures thinking involves inspecting one's beliefs, disrupting the constraints of current conceptions, and challenging assumptions about what we believe can and cannot happen (Codd et al., 2002; Ogilvy, 2006). In addition, futures thinking involves envisioning the many futures that might be realized (possible futures), the likelihood that these possible futures could occur (probable futures), and which futures are most desirable (preferable futures) (Bell, 1998). Working through these possible, probable, and preferable futures can help people anticipate and plan for what might occur, be agile in the face of change, and pursue futures they desire.

Futures thinking encompasses skills such as critical thinking, handling complexity, understanding uncertainty, making intelligent decisions for long-term sustainability, and preparing for change (Julien et al., 2018). It embraces creativity, exploration, divergent thinking, and uncertainty to seek many possible answers, differing from the type of convergent thinking that aims to predict the future and reduce uncertainty (Department of the Prime Minister and Cabinet, 2021). Futures thinking is not focused on predicting the future and determining what *will* be but instead helps us speculate about what *could* be. Futures thinking encourages us to consider what we want to occur and then work toward these preferable futures.

Inviting young people to speculate about possible futures can support them in expressing preferable futures. Engaging students in futures thinking related to the climate crisis by imagining possible sustainable and equitable energy futures serves as a rich context for working through the complexities of climate change and considering preferable futures for youth and their communities, centering their experiences and perspectives, and amplifying their voices. Students engaging in futures thinking might

- examine the history of climate change, contributing factors, existing systems and policies related to the climate crisis, relationships between climate change and their communities, and current and emerging ways people are addressing energy transitions such as solar power (Jordan et al., 2021);
- explore possible future scenarios; and
- identify with a perceived ideal future and take steps to achieve it.

Engaging in such futures thinking challenges youth to exercise agency by designing and communicating transformative socio-technical systems and environments that shape their communities. But what does this have to do with music?

Futures thinking is rife in popular culture—whether in novels such as Nnedi Okorafor's *Noor nnedi.com/books/noor t.ly/7ews*, films like Bong Joon Ho's *Snowpiercer lionsgate.com/movies/Snowpiercer t.ly/7ews*, or music such as Janelle Monae's Afro-futurist album *Dirty Computer youtube.com/playlist?list=OLAK5uy_ kqbZeag2Ai_9wn4GEH_Pfg0KfNTmqv2I0 t.ly/7ews*. Speculative fictions and futures can delight and serve as critical lenses, focusing our attention on problematic aspects of society. Educators might draw upon how artists imagine futures that reflect past and present aspects of society, providing socio-cultural and historical critiques (Bina et al., 2017), and extend this work to young people.

As Wong and Peña (2017) argue, arts engagement coupled with inquiry can support students in learning about critical issues facing society, including struggles for freedom. Working through complex socio-cultural issues for social transformation to better people's lives calls for students to be bold and imaginative about how the world could be otherwise. Hess (2021), drawing on the work of Bettina Love *bettinalove.com t.ly/7ews* and Maxine Greene *maxinegreene. org t.ly/7ews*, suggests that students can engage with music in ways where they can imagine different possible futures. We concur. The remainder of this chapter explores how educators might support youth in addressing climate change by imagining possible futures around sustainable and equitable energy systems through music.

Music and Climate Change

Discussing how artists speak to social issues, Galafassi et al. (2018) highlight how artistic practices are "heralding shifts in mindsets, opening up new political horizons and providing—sometimes even forcing—the creation of novel spaces for reflexivity and experimentation" (p. 73). By weaving art with futures thinking, artists open spaces for reflexivity and experimentation around climate change issues, inviting people to consider possibilities of more sustainable and equitable futures. Addressing climate change through the arts to imagine and speculate about a sustainable and equitable world might create the space needed for social transformation.

Artists are exploring issues around climate change and inviting speculation about what the world might be like if we do not take needed action. For instance, Paul D. Miller's (aka DJ Spooky) *The Book of Ice djspooky.com/antarctica t.ly/7ews* and corresponding music *Terra Nova: Sinfonia Antarctica djspooky.com/ terra-nova-sinfonia-antarctica t.ly/7ews* employs images, text, and sound to reflect on Antarctica and the impact of climate change. Jayda G's House dance track

Misty Knows What's Up includes samples of biologist Misty MacDuffie speaking about protecting killer whales, sparking people's curiosity about why whales are threatened. *youtu.be/_GK6V84YfZs t.ly/7ews*. In *All Good Girls Go to Hell drivestudios.tv/music/billie-eilish-good-girls-go-to-hell t.ly/7ews*, Billie Eilish comments on "man's" negative impact on the planet, warning us of a future in which our acts of violence on the planet place us beyond help. These types of warnings are hardly new. Speaking to issues in his own community, DJ Cavem addresses pollution and possible solutions, including solar power and Black power in the track, *I Can't Breathe the Air youtu.be/KC7BpqLLvYk t.ly/7ews*. In 2005, the band *Gojira* imagined a dystopian future in their song, *Global Warming listenable-records.bandcamp.com/track/global-warming t.ly/7ews*, depicting a grim world experiencing climate change. Yet, their repeated line "we will see our children growing" *genius.com/12694014 t.ly/7ews* highlights Gojira's belief in humanity, opening a speculative space to consider our ability and responsibility to act to prevent the worst climate impacts.

Though many artists explore dystopian futures, the glimmer of hope that Gojira vocalize is a starting point for other artists. For instance, in the Climate Stories Project, *climatestoriesproject.org t.ly/7ews*, Jason Davis creates and performs music intertwined with recordings of people sharing stories of their experiences with climate change and convictions for solutions, inviting contemplation and speculation about what might be and our roles in working toward sustainable futures. Hip Hop artist Xiuhtezcatl urges listeners to heal the world in his track *Broken youtu.be/LKUZJjxm9Vs t.ly/7ews* and invites people to take action in his collaboration with Jaden Smith on *Boombox Warfare youtu.be/5DA-w-_HZLo t.ly/7ews*. In a different vein, the arts movement of Solarpunk *citme.music.asu.edu/ initiatives/transforming-society/solarpunk* t.ly/7ews evokes what a solar-powered future might sound and feel like, emphasizing preferable futures. It is this turn to the possible and the invitation to explore transformational change that we explore in this chapter.

Engaging Youth in Creating Music that Speaks to Solar Futures

While popular media and music that address climate change can be powerful in sparking curiosity and concern, catalyzing dialogue, and evoking youths' imaginations around sustainability issues, actively creating media and art can open additional imaginative possibilities in addressing climate change. Inviting young people to imagine possible futures through their musical pursuits aligns with ESD as "citizenship in action" (UNESCO, 2020, p. 18) and the importance of mobilizing youth to engage in meaningful, transformative actions in their communities. In this chapter, we situate students' music creation as a medium to imagine possible futures and transformative actions in their communities. This approach differs from learning specific musical concepts or skills (though students could learn and develop musical concepts and skills by creating their music).

Our focus on linking creative musical engagement with futures thinking and STEM disciplines to address climate change builds on decades of music educators supporting students creating music. Some music educators support students in addressing social issues through music creation, and examples of educators and students exploring sustainability through music are emerging. For instance, music educator Dan Shevock (2017) describes taking students outside to listen to sounds around the nearby environment as if they were listening to a music composition and facilitated dialogue about students' experiences to promote environmental awareness. Ben Kelly (2018), a technology teacher, facilitated *Project SUSTAIN*, an international collaboration among music teachers and students, to create an album of songs reflecting the 17 United Nations Sustainable Development Goals (SDGs). Kelly noticed that older students corresponded characteristics of their music, such as genre to specific SDGs (B. Kelly, personal communication, September 8, 2022).

Since examples of music-focused projects addressing climate change in schools are sparse, educators seeking to design opportunities for students to address climate change and sustainable and equitable energy through music might look to projects and programs that broadly address social issues. For instance, Kaschub (2009) describes how students identified and analyzed existing music focusing on social justice issues to inform and inspire their creative music-making that addressed social issues that students found meaningful.

In their research of an after-school open mic program in which youth created and performed original music, Hess et al. (2019) describe how youth expressed their perspectives on the world, engaged in critical conversations through their music, and took on multiple identities through the artistic roles with which they engaged as a form of civic engagement. Teaching artists scaffolded young people's navigation of in-school and out-of-school literacies, connections between socio-cultural issues and their music, and the development of music and lyrics. This ranged from cultivating the youth's inquiry skills to modeling ethics of sharing one's work with others. Educators can link the type of engagement in these examples with ESD by encouraging students to connect their lived experiences and arts engagement to larger-scale social injustices and societal issues (Wong & Peña, 2017). This includes inviting students to imagine possible (different) futures powered by sustainable and equitable energy systems with and through music.

The Weight of Light

We offer the Weight of Light (WoL) *citme.music.asu.edu/initiatives/transforming-society/weight-of-light t.ly/7ews* project as an example of how students might engage critically with climate change and sustainable/equitable energy systems through futures thinking. Creating music or designing musical instruments supports students' development of agency and civic imagination to address the climate crisis. We describe WoL and our design process below to inspire educators seeking to address sustainability and climate change issues through creative means in their programs. WoL situates music as a creative medium to guide young people in considering complex factors around technical, energy, and social systems (National Research Council, 2013) by developing socio-technical knowledge and creating original music and musical instruments. Inspired by *The Weight of Light: A Collection of Solar Futures* (Eschrich & Miller, 2019), a set of short science fiction stories that speculate possibilities of futures powered by solar energy, WoL leverages the creative and expressive potential of music to support imagination, share stories, build future worlds, and address issues related to solar futures as a dialogue between arts and sciences (Rose et al., 2011).

Originally created as a design-based research project, WoL builds on a previous interdisciplinary collaboration involving two WoL team members, a "Solar Futures Narrative Hackathon" that enlisted adult science fiction authors, visual artists, and disciplinary experts to co-create technically grounded narratives set in a solar democracy (cf. Mitchell, 2011) accompanied by an artistic rendering of the narrative world by a visual artist, and a technical essay by a solar engineer (Eschrich & Miller, 2019). WoL builds on the hackathon by centering youth engagement and perspectives and focusing on music as the primary medium for engaging in solar futures thinking, arts-based inquiry, and creative expression.[1]

Youth worldwide, frustrated with the lack of social and political climate action and taking action themselves, also inspired WoL. Embracing young people's role in addressing climate change, WoL explores how futures thinking might foster youth STEAM agency, helping youth navigate the unknown space of possibility and opportunity that lies ahead for transformations of energy systems. WoL invites young people to imagine possible sustainable and equitable energy systems in relation to their communities and society through designing solar-powered instruments or creating music. It, therefore, can increase interest in and awareness of energy challenges (Jordan et al., 2021) while supporting youth agency in engaging art and science in ways that matter in their lives, communities, and society. WoL encourages students to be bold in envisioning possibilities for the future and courageous in imagining how democratic societies might work to bring about preferred energy systems.

At the center of WoL are questions such as: What would a world powered entirely by solar energy look [and sound] like? What would it be like to live in such a world? (Eschrich & Miller, 2019). How might youth engage in creative musical expression to convey their concerns about energy and its role in climate change? WoL also explores questions such as: How do the arts and sciences come together to imagine possible, probable, and preferable futures? How might we facilitate ways for people across the arts, sciences, and education to collaborate? What social infrastructures support equitable collaboration among young people, artists, scientists, and arts and science educators?

While these questions informed our research and development of WoL, we encourage educators to consider their wonderments and their students' curiosities about combining science and arts with working through the complexities of climate change and imagining possible solar futures. Finally, we describe WoL and our design process in the spirit of offering an approach to address climate change through combining science with music without being prescriptive and inspiring readers to adapt this project or develop their related projects with their learning communities.

Our design process included:

1 Agreeing on project goals.
2 Identifying teachers open to addressing music creation, futures thinking, and climate change in their program.
3 Designing a project structure.
4 Curating and designing resources and prompts to foster futures thinking, spark curiosity about solar power, and support creative engagement.
5 Developing a design brief to situate the project.

It is important to note that disruptions of the COVID-19 pandemic thwarted our original plan to have music and science educators facilitate WoL collaboratively. Though the project occurred in music education contexts, we envision WoL in its ideal form occurring across multiple types of classes and encourage educators interested in related projects to engage interdisciplinarily in their programs.

Goals, Objectives, and Outcomes

Prior to starting the project with students, WoL team members discussed and agreed on project goals. We established WoL as design-based research, which meant embracing an ethic of curiosity and exploration regarding what might occur rather than planning for predetermined outcomes. We had the overarching goal of learning from the project to better understand the potential of arts-based STEM/STEAM engagement and learning in helping students consider the complexities around climate change and imagine possible futures for themselves and their communities. A shared set of principles also informed the project design. For instance, we were interested in the potential for students to consider positive solar futures while keeping space for those interested in expressing themselves and their perspectives through more negative portrayals of what might be. We were also interested in encouraging students to consider their local communities and socio-cultural issues about climate change and sustainable and equitable energy systems. The goal of supporting student agency was also core to the project. Though we did not articulate particular student learning outcomes or objectives, educators can layer more specific curricular objectives or outcomes into projects such as WoL.

Collaborating

The complexity of climate change invites interdisciplinary collaboration. While educators can facilitate music-focused projects addressing climate change individually, collaborating on an interdisciplinary team is ideal. Our team involved two high school music educators Bill Singer and David Ruiz (pseudonyms), university educators across disciplines with expertise in fields such as futures thinking, STEM/STEAM education, sustainable and equitable energy systems, design, creative writing, and music education; graduate researchers; and designers and architects who specialize in linking design, art, education, and renewable energy from the Land Art Generator (LAGI) *landartgenerator.org/index.html t.ly/7ews*. Our partnership with LAGI was integral in developing the design brief and generating ways to catalyze thinking around the design of possible solar futures. While some of our team members had worked with one another on previous projects, WoL was the first time all team members collaborated as a transdisciplinary team.

We selected the music educators involved in WoL because of their open-mindedness, flexibility to adapt to the project's emergent nature, learner-centered orientation to their programs, and existing working relationships from prior collaborations. Though we did not include K-12 students on the WoL team, educators might consider including students as part of the planning process to ensure that youth perspectives are present to inform such projects (Cahill & Dadvand, 2018). We relied on Bill and David's familiarity and ongoing dialogue with their students to account for youth perspectives and to ensure that WoL would be meaningful and appropriate for their learning communities. Unfortunately, our plan for a local popular musician to support students' creative work was thwarted when schools moved online during the COVID-19 pandemic. However, we encourage educators to consider partnering with local artists, particularly those experienced working with youth in varied educational settings. To maintain forward movement with our planning, we met weekly for several months and invited additional collaborators to dialogue with us as we designed WoL.

We include these details regarding our team since choosing collaborators and working as a team are key to interdisciplinary projects. Projects such as WoL are excellent opportunities for people with shared interests in climate change education to develop collaborative relationships and partnerships. Who might you collaborate with to support students' imagination of possible solar futures through music and other arts?

Designing a Project Structure

While we approached WoL with an open and exploratory ethic, we provided a flexible structure for teachers and students. We designed the project to start with optional resources, prompts, and activities *citme.music.asu.edu/initiatives/ transforming-society/weight-of-light/wol-activities t.ly/7ews* for music classes to try, leading to a brief that situated the project with a design scenario to address. Starting with shorter-term activities gave teachers the flexibility to begin WoL without disrupting their existing program plans. We designed the prompts and activities to spark students' curiosity and interest in solar energy and climate change through low-stakes exploration and play. While the prompts and activities could serve as one-off explorations of climate change and sustainable and equitable energy systems, collectively, the prompts and activities functioned as scaffolding for the type of engagement and thinking to be applied and deepened through the core project. This type of scaffolding is critical to more open-ended projects. Consider how you might structure projects in ways for students to feel supported when encountering openness and ambiguity.

The core component of the project involved students creating music or instruments in response to the design brief that LAGI created with our team and participating music teachers. Students' efforts to respond to the brief such as their music creation, instrument design, reflections, and discussions occurred in a style similar to project-based learning (Larmer et al., 2015). The brief provided context and criteria yet was open-ended to foster multiple approaches for students to imagine and create solar futures through music, differing from a more highly structured and sequenced lesson and project plans. Thus, WoL aligned with the principles of ESD in situating sustainability through a learner-centered approach in which students engaged in inquiry and project-based learning rather than focusing primarily on memorizing facts about sustainability or climate change or applying their knowledge in a standardized manner.

Curating and Designing Resources

We designed WoL to support students drawing upon their own perspectives, lived experiences, and funds of knowledge (Bylica, 2022; Hess, 2021; Moll et al., 1992; Wong & Peña, 2017) along with information about solar power.

In collaborating with local music educators, we considered the importance of providing creative prompts and scaffolds to inspire or catalyze students' creative work and imaginations. Before students started WoL, we designed and curated resources such as websites and media that explained concepts related to solar power, musicians who address climate change through their art, and specific music examples. Readers can access these and related resources via the WoL project website *citme.music.asu.edu/initiatives/transforming-society/weight-of-light t.ly/7ews*.

For instance, we shared examples *citme.music.asu.edu/initiatives/transforming-society/weight-of-light t.ly/7ews* of artists and resources that addressed solar power explicitly, such as the work *Beware of The Dandelions emergencemedia.org/pages/beware-of-the-dandelions t.ly/7ews*, the Solarpunk zine, *Optopia issuu.com/optopia t.ly/7ews*, the organization Artists and Climate Change *artistsandclimatechange. com t.ly/7ews*, a set of web-based music creation apps *evantobias.net/web-based-music-apps t.ly/7ews*, the website for The Quantum Energy and Sustainable Solar Technologies (QESST) *qesst.org t.ly/7ews*, and information on the Crescent Dunes Solar Energy Project *en.wikipedia.org/wiki/Crescent_Dunes_Solar_Energy_Project t.ly/7ews*. Then, as a team, we read *For the Snake of Power* (Cooper, 2018), a short story from *The Weight of Light: A Collection of Solar Futures* (Eschrich & Miller, 2019), and invited Bill and David to include *For the Snake of Power* as a resource for students. The story offered a unique way of thinking through issues around solar power and energy transitions that differed from scientific studies or sets of facts about photovoltaics or climate change.

Additionally, our collaborators from LAGI created a slide deck of inspirational and informational materials demonstrating the possibilities of solar energy and design *citme.music.asu.edu/initiatives/transforming-society/weight-of-light/wol-activities t.ly/7ews*, extending beyond the solar panels that were more familiar to students. We also worked with LAGI to develop a design brief to situate and guide WoL. Throughout WoL, students answered prompts that our team and the music teachers developed to support reflection *citme.music.asu.edu/initiatives/transforming-society/weight-of-light/wol-reflection-prompts t.ly/7ews*. Bill and David also curated and developed their resources for students. For instance, Bill shared detailed instructions, processes, and resources for designing instruments with students.

Developing a Design Brief as a Context for Engagement and Learning

After students initially explored solar power and creative music engagement through the activities and prompts, Bill and David introduced the design brief to frame the project. A design brief is "the statement of the problem: its goals, parameters, constraints, and component parts, all of which have tangible and intangible aspects" (Pendleton-Jullian & Brown, 2018, p. 94). The "problem" to address in WoL is to imagine a future of one's community powered by sustainable and equitable energy systems rather than fossil fuels and to express what that solar future might be like. Addressing climate change and the cultural, historical, social, scientific, and technical issues through the arts is a complex problem. Pendleton-Jullian and Brown (2018) differentiate complex problems, which are "embedded within human contexts that organize themselves through changing social, political, economic, and cultural belief systems" (p. 45), from complicated problems, which are perceived as "only scientific or technical in nature, removed from the contexts to which [people] are responding" and "can be solved through straightforward, scientific and engineering design methods" (p. 45).

Instead of thinking of a design problem as something to solve, we can consider design problems as "opportunities for working on the questions, puzzles, and enigmas that are inherent in human existence" (Pendleton-Jullian & Brown, 2018, p. 26). The goal with WoL was not for students to solve the complex problem of climate change, but rather to work through the questions, puzzles, and enigmas of climate change and energy transitions through creative music engagement in ways that were meaningful to their lives and the lives of those in their communities while potentially developing interest in related STEM fields.

Our team developed the WoL design brief iteratively in collaboration with LAGI and the participating music teachers. After LAGI developed the brief for WoL, we had initial conversations and made adjustments. We then brought Bill and David into dialogue with the team to provide their perspectives on the brief with their music programs, students, and communities in mind. Finally, the teachers provided feedback that informed collaborative revisions of the brief until we agreed upon a version that would work well with students in both music programs. Below, we include a brief excerpt of the design brief that students engaged with during WoL.

> The year is **2045**. The previous two decades have brought many challenges as the world transitioned from fossil fuels to renewable and clean forms of energy. As a celebration of the great energy transition your city has put out a call for a new solar powered music experience. Your challenge is to tell the story of how people experience music and art in a 100% solar powered future and how music and art were so important during the years of change (2025–2045). As you consider your response to the city's creative call for proposals, consider how life in this new solar future has transformed human culture, the ways we relate to one another, the way we work, our civic participation, art-making, or recreation. Working either individually or as part of a team you will present your vision of a solar powered future for the year 2045.

We invite you to read and engage with the design brief in its entirety via the WoL project website *citme.music.asu.edu/initiatives/transforming-society/weight-of-light/wol-designbrief t.ly/7ews*. Consider how the brief invites imaginative thinking, addresses climate change, catalyzes futures thinking, and situates music as a medium for working through complex problems. We also encourage you to consider what a similar brief might be like for your learning community. Though educators can adopt the WoL brief for their own classrooms, it is important to create design briefs that are customized and meaningful to the learning communities who are engaging in these types of projects.

By situating climate change, sustainable energy transitions, and related socio-cultural and technical issues as wicked problems to address through music engagement, the brief provides a catalyst to spur students' imaginations, a structure to organize the project, parameters to guide students' engagement, and instructions about submitting project work. In this way, the brief serves as a context for students' engagement and learning, with educators supporting students by facilitating and scaffolding along the way. Yet, it is how students address the questions, puzzles, and enigmas of the brief through their music engagement, reflections, and dialogue where their imaginations of possible solar futures come alive.

Expanding the Brief: Project Work

Pendleton-Jullian and Brown (2018) describe a process of expanding the brief in which designers, or students, ask questions, address constraints, take in information, and work through the problem. As they explain, "The expanded brief is about moving the process forward through perpetual inquiry in areas that may not be the most obvious to the project at hand. Continually expanding the brief leads to a deep empathy with the problem and with the context in which the problem is embedded" (p. 95).

In WoL, the design brief presented an opportunity for students to work through the wicked problem of climate change through music by challenging them to tell the story of how people experience music and art in a 100% solar-powered future. Students then created music about transitioning from fossil fuels to sustainable and equitable energy systems. During our research of WoL, students expanded the brief by imagining possible solar futures through music in two primary ways (1) designing and creating solar-powered instruments and (2) creating original music. Both music teachers supported students' agency in determining what and how they would create. Some students worked individually, while others worked in groups. Students sometimes collaborated as an entire class. Detailing students' musical processes and music *citme.music. asu.edu/initiatives/transforming-society/weight-of-light/wol-student-project-examples t.ly/7ews* is beyond this chapter's scope. However, we will discuss several of

Bill's and David's pedagogical approaches to facilitate students' inquiry and creation.

Making Space for Reflection and Dialogue

While the sonic aspects of students' music creations are important, music educators who facilitate projects similar to WoL may need to adjust the focus of their feedback and criteria for success. For example, students' original music addressing possible sustainable and equitable energy systems or other aspects of climate change opens space for dialogue or prompts new ways of thinking about energy, more so than how the music follows a particular form or chord progression. This means including reflection and dialogue as part of the project itself.

Bylica (2022) highlights this shift of focus from the composition and the act of composing to "the creation of space for that composition to be repeatedly heard, discussed, and problematized" (p. 17) to support student dialogue. Bylica also found that facilitating multiple contexts for reflection and dialogue, ranging from individual and small group reflection to entire class discussion, can help students gain trust and share their perspectives. Bill and David provided multiple formats for students to reflect and dialogue, such as individual Flipgrid videos, small group and one-to-one conversations in physical space and via Zoom, and whole-class discussions.

For example, during a class that met on Zoom, David asked students to share what they remembered or what was interesting about solar futures and helping people improve the world via their preference of chat or verbally. He also supported and affirmed students by elaborating on their responses that encouraged others to respond. David also maintained a casual and relaxed pace that held space for students to gather their thoughts, such as joking around about starting a Thrash-cumbia band while encouraging students to share their future music genres and how they might sound. David's facilitation approaches and openness to students' perspectives and interests throughout WoL supported their engagement with new and atypical concepts in music programs.

Bill took a more structured approach to facilitating student reflection and dialogue, such as employing a four-stage process from his work with e4USA (Engineering for US All, 2022) that included questions corresponding to the processes of Inspiration, Planning, Creation, and Evaluation. For example, during the Inspiration phase, Bill asked students to identify the inspiration of their music and "why it appeals to you (or others) based on motivation/purpose." He further scaffolded questions with prompts guiding students' answers, such as "I really like the music of ____ because ____ and want to see if I can create something similar." Bill also required students to share weekly video updates on their project progress, though responses were typically brief with little detail. This may have related to their overall experiences of school and life throughout the COVID-19 pandemic.

Approaches to Supporting Students Create Original Music

Providing students with space to create music in ways that are most meaningful to them, including working on their preferred forms of musicianship, can support their creative and expressive engagement. Bill and David scaffolded students' pursuit of their desired musical trajectories in terms of music and instrument creation by keeping students' options open and not requiring them to meet specific musical criteria that are sometimes typical in music creation exercises in K–12 programs. Though the project stipulated parameters relating to the theme of a solar-powered future, the openness in which students could create led to diverse creative approaches and sonic possibilities.

Music educators should reflect on how the criteria they require in music creation projects might constrain or force students to fit their art and ideas into conventional Eurocentric arts forms or practices (Wong & Peña, 2017). This might mean avoiding or reducing criteria that specify particular rhythm or pitch patterns, genres, form, or other aspects of a music curriculum. As Wiggins (2007) reminds us, educators play an important role in supporting or hindering students' agency in facilitating music creation.

However, educators are also responsible for scaffolding students' engagement and learning when appropriate, including sharing prompts and resources that offer starting points when students are overwhelmed with openness or desire assistance in their creative process. For instance, students' exploration of fictional music genres *citme.music.asu.edu/initiatives/transforming-society/weight-of-light/wol-activities t.ly/7ews* in WoL provided a playful opportunity to speculate about what music of the future might sound like and catalyzed imaginative thinking about what students might create. The openness of creating music indicative of "Viral Sky Shanties" or "Thawpop" (two WoL fictional music genres) differs from requiring students to create music that includes particular rhythm patterns or follows specific harmonic structures. Educators unfamiliar or uncomfortable with supporting students in creating music in ways that extend beyond Western classical music paradigms might collaborate with artists more familiar with the kinds of music and musical practices that students enjoy (Gage et al., 2020; Hess et al., 2019).

Design Considerations

When planning projects such as WoL, consider how open-ended the project should be. For instance, projects can serve as contexts for students to work through issues around climate change, solar energy, and possible futures without having specific learning outcomes predetermined. This aligns with Eisner's (2002) notion of expressive outcomes, which emerge from engagement rather than being decided before starting a project. Eisner explains that expressive outcomes "are the consequences of curriculum activities that are intentionally planned to provide a fertile field for personal purposing and experience" (pp. 118–119). Alternatively, educators can design projects that address climate change through musical engagement with specific learning objectives or outcomes aligned with one's curriculum.

For instance, science educators might design a project similar to WoL so that students learn concepts related to solar energy or aspects of Next Generation Science Standards. Music educators might include parameters that address concepts related to creating or performing music. Educators might also draw upon ESD to foster students' development of systems thinking and informed decision-making that support agentive and transformational engagement with their communities. While Bill had specific creative processes he planned for students to learn, David focused on whatever emerged from students' engagement and discussion. Both educators, however, maintained an open and flexible approach to how students made meaning of the project.

Educators collaborating in interdisciplinary teams can determine what goals, objectives, or outcomes to emphasize or de-emphasize across classes and disciplines depending on one's expertise and the learning context. Whether drawing upon curriculum, standards, or the expressive outcomes of students' project work, we should be intentional about how we design projects addressing climate change and sustainability and facilitate students' engagement.

Possibilities Ahead

Addressing the wicked problem of climate change calls for us to consider the broadest spectrum of approaches across schools and other settings, including those that are arts-based and informed by STEM disciplines. The WoL project is one example of how young people can combine science and music to imagine possible solar futures for themselves and their communities. Situating the science of climate change and sustainable and equitable energy transitions in students' musical engagement has the potential to amplify youth voices through music and dialogue that may play a decisive role in contributing to needed change. We encourage educators to design and facilitate similar projects with colleagues and students to catalyze imagination of what could be and to spark needed transformational change.

Note

1   While this chapter focuses on the music-specific aspects of WoL, the project also included a STEM program creating futures narratives that explored the possibilities of agrivoltaics.

References

Beach, R., Share, J., & Webb, A. (2017). *Teaching climate change to adolescents: Reading, writing, and making a difference*. Routledge.

Bell, W. (1998). Making people responsible: The possible, the probable, and the preferable. *American Behavioral Scientist, 42*(3), 323–339. doi:10/fwdw8f

Bina, O., Mateus, S., Pereira, L., & Caffa, A. (2017). The future imagined: Exploring fiction as a means of reflecting on today's grand societal challenges and tomorrow's options. *Futures, 86*, 166–184. doi:10/gft97x

Bresler, L. (1995). The subservient, co-equal, affective, and social integration styles and their implications for the arts. *Arts Education Policy Review, 96*(5), 31–37. doi:10.1080/10632913.1995.9934564

Bylica, K. (2022). "You can only choose from the things you know": Engaging with students' dark and politicized funds of knowledge in the music classroom. *Education and Urban Society*, (onlinefirst), 1–24. doi:10.1177/00131245221087995

Cahill, H., & Dadvand, B. (2018). Re-conceptualising youth participation: A framework to inform action. *Children and Youth Services Review, 95*, 243–253. doi:10/gfkjtn

Codd, J., Brown, M., Clark, J., McPherson, J., O'Neill, H., O'Neill, J., Waitere-Ang, H., & Zepke, N. (2002). Review of future-focused research on teaching and learning. *Wellington: Ministry of Education*. https://www.educationcounts.govt.nz/publications/schooling2/curriculum/5331

Cooper, B. (2018). For the snake of power. In J. Eschrich & C. A. Miller (Eds.). *The weight of light: A collection of solar futures* (pp. 43–59). Arizona State University Press.

Department of the Prime Minister and Cabinet. (2021, November 23). Futures thinking. https://dpmc.govt.nz/our-programmes/policy-project/policy-methods-toolbox/futures-thinking

Drehobl, A., & Ross, L. (2016). *Lifting the high energy burden in America's largest cities: How energy efficiency can improve low income and underserved communities*. https://www.aceee.org/research-report/u1602

Eisner, E. (2002). *The educational imagination. On the design and evaluation of school programs* (3rd ed.). Merrill Prentice Hall.

Engineering for US All. (2022). Engineering For US All. https://e4usa.org/

Eschrich, J., & Miller, C. A. (2019). *The weight of light: A collection of solar futures*. Arizona State University, Center for Science and the Imagination.

Fisher, E., & Mahajan, R. L. (2010). Embedding the humanities in engineering: Art, dialogue, and a laboratory. *Trading Zones and Interactional Expertise: Creating New Kinds of Collaboration*. doi:10/gqshpx

Gage, N., Low, B., & Reyes, F. L. (2020). Listen to the tastemakers: Building an urban arts high school music curriculum. *Research Studies in Music Education, 42*(1), 19–36. doi:10/gqr8p9

Galafassi, D., Kagan, S., Milkoreit, M., Heras, M., Bilodeau, C., Bourke, S. J., Merrie, A., Guerrero, L., Pétursdóttir, G., & Tàbara, J. D. (2018). 'Raising the temperature': The arts on a warming planet. *Current Opinion in Environmental Sustainability, 31,* 71–79. doi:10/gc2fpj

Giangrande, N., White, R. M., East, M., Jackson, R., Clarke, T., Saloff Coste, M., & Penha-Lopes, G. (2019). A competency framework to assess and activate education for sustainable development: Addressing the UN sustainable development goals 4.7 challenge. *Sustainability, 11*(10), 2832. doi:10/ggmjpp

Hess, J. (2021). Musicking a different possible future: The role of music in imagination. *Music Education Research, 23*(2), 270–285. doi:10/gjk9

Hess, J., Watson, V. W., & Deroo, M. R. (2019). "Show some love": Youth and teaching artists enacting literary presence and musical presence in an after-school literacy-and-songwriting class. *Teachers College Record, 121*(5), 1–44. doi:10/gqr8qb

Intergovernmental Panel on Climate Change. (2022). Climate change 2022: Impacts, adaptation and vulnerability. In H.-O. Pörtner, D.C. Roberts, M. Tignor, E.S. Poloczanska, K. Mintenbeck, A. Alegría, M. Craig, S. Langsdorf, S. Löschke, V. Möller, A. Okem, B. Rama (Eds.), *Contribution of working group II to the sixth assessment report of the intergovernmental panel on climate change.* Cambridge University Press. doi:10.1017/9781009325844

Jordan, M., Bernier, J., & Zuiker, S. (2021). The future is open and shapable: Using solar speculative fiction to foster learner agency. *Literacy Research: Theory, Method, and Practice, 70*(1), 309–329. doi:10/gqr8nf

Julien, M.-P., Chalmeau, R., Vergnolle Mainar, C., & Léna, J.-Y. (2018). An innovative framework for encouraging future thinking in ESD: A case study in a French school. *Futures, 101,* 26–35. doi:10/gdzm53

Kaschub, M. (2009). Critical pedagogy for creative artists: Inviting young composers to engage in artistic social action. In E. Gould, J. Countryman, C. Morton, & L. S. Rose (Eds.), *Exploring social justice: How music education might matter* (pp. 274–291). Canadian Music Educators Association.

Kelly, B. (2018, October 3). Digital music-making: Cross-cultural collaboration. *Soundtrap.* https://edu.soundtrap.com/digital-music-making-cross-cultural-collaboration/

Kysilka, M. L. (1998). Understanding integrated curriculum. *The Curriculum Journal, 9*(2), 197–209. doi:10/dmvg63

Larmer, J., Mergendoller, J., & Boss, S. (2015). *Setting the standard for project based learning.* ASCD.

Littrell, M. K., Tayne, K., Okochi, C., Leckey, E., Gold, A. U., & Lynds, S. (2020). Student perspectives on climate change through place-based filmmaking. *Environmental Education Research, 26*(4), 594–610. doi:10/gpfqdz

Miller, R. (2003). *Where schools might fit in a future learning society.* Incorporated Association of Registered Teachers of Victoria.

Mitchell, T. (2011). *Carbon democracy: Political power in the age of oil.* Verso.

Moll, L. C., Amanti, C., Neff, D., & Gonzalez, N. (1992). Funds of knowledge for teaching: Using a qualitative approach to connect homes and classrooms. *Theory into Practice, 31*(2), 132–141. doi:cbg3gt

National Research Council. (2013). *Next generation science standards: For states, by states.* The National Academies Press. doi:10.17226/18290.

Ogilvy, J. (2006). *Education in the information age: Scenarios, equity and equality.* In *Think scenarios, rethink education*, OECD Publishing. doi:10.1787/9789264023642-3-en

Pendleton-Jullian, A. M., & Brown, J. S. (2018). *Design unbound: Designing for emergence in a white water world* (Vol. 1). MIT Press.

Rose, C., Smith, B., & Ryan, S. (2011). *Report to the National Science Foundation on bridging STEM to STEAM: Building new frameworks for art/science pedagogy.* Rhode Island School District Press Release. https://sead.viz.tamu.edu/about/pdf/STEAM.pdf

Shevock, D. J. (2017). *Eco-literate music pedagogy: Routledge new directions in music education.* Routledge.

United Nations Educational Scientific and Cultural Organizations. (2014, June 13). Education for sustainable development. UNESCO. https://en.unesco.org/partnerships/partnering/education-sustainable-development

United Nations Educational Scientific and Cultural Organizations. (2020). *Education for sustainable development: A roadmap.* https://unesdoc.unesco.org/ark:/48223/pf0000374802

United Nations Educational Scientific and Cultural Organizations. (2022, June 9). *What you need to know about education for sustainable development | UNESCO.* https://www.unesco.org/en/education/sustainable-development/need-know

United Nations Educational Scientific and Cultural Organizations & Mahatma Gandhi Institute of Education for Peace and Sustainable Development. (2017). *Textbooks for Sustainable Development: A Guide to Embedding.* UNESCO-MGIEP. https://unesdoc.unesco.org/ark:/48223/pf0000259932

Wiggins, J. (2007). Compositional process in music. In L. Bresler (Ed.), *International handbook of research in arts education* (pp. 453–476). Springer.

Wong, C., & Peña, C. (2017). Policing and performing culture: Rethinking "culture" and the role of the arts in culturally sustaining pedagogies. In D. Paris & S. Alim (Eds.), *Culturally sustaining pedagogies: Teaching and learning for justice in a changing world* (pp. 117–138). Teachers College Press.

Zraick, K. (2019, November, 1). Greta Thunberg, after pointed UN speech, faces attacks from the Right. *New York Times.* https://www.nytimes.com/2019/09/24/climate/greta-thunberg-un.html?smid=url-share

8

FOSTERING PROACTIVE ECOLOGICAL IDENTITY OF YOUTH THROUGH SOCIAL MEDIA

Nataliia Goshylyk

The debates over human-nature relations and ecological awareness have been one of the central concerns in contemporary Sciences and Humanities in the 20th and especially 21st centuries. The rapid technological and economic growth after World War II, the constant need to exploit natural resources, and the consequential negative impact on the environment have become the primary characteristics of the Anthropocene, a concept introduced "to name and explore the human impacts on the Earth System" (Sklair, 2021, p. 5). This growth has been analyzed predominantly from the point of view of the consequences the human beings' activities have led to and the extraordinary scale, scope, and magnitude of this impact (Hoffman & Jennings, 2018; Howe, 2019; Lewis & Maslin, 2018; Sklair, 2021). Bearing in mind that the Anthropocene needs to be deframed and reframed to include diverse social values, needs, and wants (Castree, 2021), this strand shaped the ecological agenda of the globe and reinforced the need for raising awareness and acting for the good of the planet.

The 21st century has created endless opportunities for personal and institutional networking, expanding ground activism, and engaging people, who otherwise would not learn about the needs, initiatives, and actions needed. This involvement has been dynamic in the social sphere, especially regarding human rights and education, having a strong focus on environmental issues. In addition, the possibilities for displaying and performing ecological identity grew with the spread of social media where everyone can become a journalist or an influencer for at least a small number of people.

Ecological activism and social media involvement have been researched from various perspectives, like ecological message circulation, online strategies, and successful communication practices (Bloomfield & Tillery, 2019;

DOI: 10.4324/9781003335276-11

Comfort & Hester, 2019; Lee et al., 2018). A recently detailed overview by Liu and Kim (2022) shows that over the past years, researchers have documented the important role of social media and contributed to understanding the human dimensions of ecological discourse. Young people's engagement has been analyzed through reports on ecological campaigns (Napawan et al., 2017), classroom practices of stories-we-live-by (Damico et al., 2020), multimedia narratives (Smith et al., 2021), digital tools (Beach & Smith, 2020), emojis (Field, 2021), the role of metaphors in climate change talk of young people (Deignan & Semino, 2022), and other action-related issues (Boulianne et al., 2020; Boulianne & Theocharis, 2020; Tyson et al., 2021). This chapter presents an overview of how social media's underrepresented discursive and systemic features foster young people's needs and abilities to set a global decolonizing environmental agenda.

Social Constructivism Underpinnings of Proactive Ecological Identity

Ecological identity is the socially constructed category built upon objective reality and subjectively derived meanings of the dual character "that makes its 'reality sui generis'" (Berger & Luckmann, 1991, p. 30). Ecological identity could be schematically described as an individually defined extension of "our sense of self in relationship to nature" (Thomashow, 1995, p. 3). Bringing ecological identity into a broader social framework involves perceiving identity as situationally motivated and achieved (Bauman, 2000, p. 1). Social constructivists presume that identity is not ready-made, given, or formed innately. Rather, enacting identity depends on interactional occasions (De Fina et al., 2006) of multilayered character, resulting from the negotiation and entextualization process (Bauman & Briggs, 1990).

Identity is a discourse entity existing, developing, being influenced by, and influencing discourse itself. Identity is generated in discourse and generating discourse at the same time. From this perspective, discourse is powerfully creating and enforcing identity, summing up daily interactions that create and recreate us continually (R. T. Lakoff, 2006). These interactions are one of the milestones of identity performing. However, at the same time, the frequent contact occasions problematize the concept of identity itself and highlight the need to understand the relations of people and community, social norms, and practices that delineate, describe, and reveal community involvement (De Fina, 2006, p. 351).

If we are talking about ecological identity, then encounters happen between human beings and between human beings and nature. Thus, these contacts are much more frequent, natural, and happening nonstop, as one can be separated from other people but cannot be separated from nature or at least some elements of nature. So, nature becomes "the other" that is not separable from a human being, and awareness of its status and role is vital. The relations between a human being and nature can differ and depend on external and internal circumstances. In this regard, it's essential to understand one's relation to nature and one's environmental role in a wider social context, which is a part of a society member's social knowledge and involvement. Society is not a necessary condition for building relations with nature but being a society member presupposes an awareness of the society's normative framework concerning nature. Thus, ecological identity is not a ready-made phenomenon; it instead requires a social awareness of environmental issues and engagement in actions to tackle them.

Critical awareness of the world around people has three stages:

1 Noticing the world around them.
2 Seeing the influence of the world on one's life/being.
3 Understanding one's influence on the surrounding environment.

However, without this awareness, conscious actions regarding nature wouldn't be possible. These are prerequisites of performativity as the capacity to act. It is "not a singular act, but a repetition and ritual, which achieves its effects through its naturalization in the context of a body, understood, in part, as a culturally sustained temporal duration" (Butler, 1990, p. xv). Thus, identity, being of a performative character, requires actions and continuousness.

Proactive ecological identity builds on acting in the present to mitigate existing ecological problems in anticipation of future ecological challenges, needs, or changes. Proactive identity combines temporal and activity domains and builds upon personal involvement. Parker et al. (2010) view proactivity as enabling things, anticipating and preventing problems, and seizing opportunities. It is about self-initiated actions bringing change to alter the present and achieve a different future. The notion of proactivity originates from psychology with a consensus on the importance of being future-focused, change-orientated, and goal-driven (for a detailed overview, see Parker et al., 2010). From the perspective of proactive vs. reactive instructional approaches, it emphasizes proactive teaching activity for emotional well-being in the classroom (Page & Page, 2000). People do not have to wait to learn what actions to take and what needs to be done at a specific moment in a specific area. Rather, they should take the initiative to seize opportunities and actively shape the discourse, performing their identity.

The Contemporary Information Landscape

The relations between people and modern media, particularly young people and modern media, have been extensively researched from various perspectives.

The second decade of the 21st century has witnessed a drastic increase in the number of new media outlets and skyrocketing social media consumption. New media are "those that are associated with the post-broadcast era of interactive or participatory communication using networked, digital, online affordances" (Hartley et al., 2013, p. 3). Social media, unlike traditional media, are very popular with young people. This potential should be used to reinforce people's knowledge and well-being and advance the planet's good.

The capability of social media to be used for setting the environmental agenda and sharing it with the public is enormous. Not only do young people engage in more than three hours of screen time (Bergmann et al., 2022), but they also generate content that can be easily accessible to other young people via online communication. Unlike the case of broadcast media, the audience of a new type of content might be limited, leaving the impression of "many performing for the few" (Albrechtslund, 2013, p. 312). However, these are not determining restrictions because personal engagement and interaction on community levels compensate for any possible limitations.

From the perspective of information circulating, the advancement of the new media introduced a social media user who is engaging in online communication and responding to information triggers. At the same time, the user is an independent player in the system, capable of producing their messages and narratives. While traditionally, information was coming from the top to bottom, i.e., from elites (governmental officials, statespersons, etc.) through official mainstream media to the public, digitally enabled "pump-valves" in the flow of socio-political information revised the cascading activation model (Entman & Usher, 2018) and reshaped the information landscape globally.

Governmental officials have habitually set the ecological agenda, as in: "The Biden plan for a Clean Energy Revolution and Environmental Justice" (Biden, 2019), "Decision (EU) 2022/591 of the European Parliament and of the Council of 6 April 2022 on a General Union Environment Action Programme to 2030" (European Parliament & Council of the European Union, 2022), "The European Green Deal" (European Commission, 2019), "Transforming our world: the 2030 Agenda for Sustainable Development" (the United Nations, 2015).

Global ecological problems have also occupied the front pages of leading newspapers with various success and frequency. The analysis of the Corpus of Contemporary American English (more than one billion words from 1990 to 2019) and its media section shows explicitly that the word combination "climate change" has been used more frequently in recent years (data include lemmas from 1990 up to 2019). This proves that the mainstream media resonate with the ecological tensions in society and respond accordingly (see Figure 8.1).

Social media are diversifying the flow of information and, despite enabling the spread of fakes, mis-, dis-, malinformation, trolls, and bots engagement, remain the most affordable and accessible tool for global and local agenda-setting.

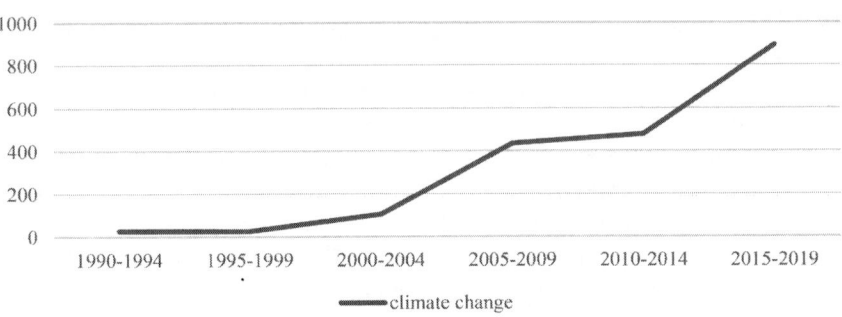

FIGURE 8.1   "Climate Change" Representation in COCA

In addition, audiences contribute to media dynamics by interacting on social media platforms (Landis & Allen, 2022). However, they also become the authors of messages and posts, thus performing their identity online and expressing themselves, which is one of the biggest attractions for young people.

The agency of young people is reinforced and facilitated by social media. Numerous engagement possibilities amplify their voice and strengthen their interaction potential. In addition, social media users have endless chances to try out their online communication patterns and choose the most appropriate one. In terms of using this creative potential of social media to reinforce young people's agency and apply their knowledge, it's worth mentioning that people use a set of cognitive skills on social media.

In terms of the revisited Bloom's taxonomy, the cognitive processes involved in using social media include focusing on *remembering* (recognizing and recalling), *understanding* (interpreting, exemplifying, classifying, summarizing, inferring, comparing, explaining), *applying* (executing, implementing), *analyzing* (differentiating, organizing, attributing), *evaluating* (checking, critiquing), and *creating* (generating, planning, producing) and employing the *factual*, *conceptual*, *procedural* and *metacognitive* types of knowledge (Bloom & Anderson, 2014). Objectives reached as the result of understanding and creating are usually considered the most important outcomes of education (Krathwohl, 2002). Media literacy skills, an inevitable prerequisite of being a conscious and cognate social media user, are based on understanding and evaluating information. For active users, social media are 24/7 platforms for testing all types of cognitive skills, the creative ones in particular. Passive social media users may remember, understand, apply, analyze, and evaluate, leaving out the production of self-generated content. In contrast, active social media users, in performing their proactive identity, show a great level of understanding, evaluating, and generating content online to share it with the public, expanding the ecologically sensitive type of discourse.

Social media attract young people to promote more sustainable understandings of living well and highlight their good life narratives (Loukianov et al., 2020). They also share their emotions discussing global warming and climate change on Twitter with reasonable reactions (Cameron et al., 2022). They may also connect their own lives to the issues related to climate change using photographing and talking about food in social media posts (Napawan et al., 2017). Constructing specific inventories of identities online builds on different types of narratives emerging in different interactional contexts (De Fina, 2006), social media being this type of context. Young people are engaging in finding new getaways to approach ecological hazards and respond to them actively and efficiently.

Social Media Decolonizing Affordances for Ecological Activism

Nowadays, ecological hazards occur all over the globe, and unfortunately, not a single place on the Earth is protected from the negative influence of the environmental crisis. Environmental problems are pollution, global warming, ozone layer depletion, acid rain, natural resource depletion, overpopulation, waste disposal, deforestation, and biodiversity loss (Singh & Singh, 2016). These issues are not immediately traced in every part of the world but are due to well-advanced globalization, with the short-term, and long-term influence of these threats is seen everywhere. Christoff and Eckersley (2013) claim that the contemporary form of globalization is "more environmentally destructive than any previous phase of globalization and, like a snake swallowing its own tail, is ultimately self-destructive" (p. 9).

Several global ecological campaigns were initiated by individuals, sustained on social media, and featured on all media types. Greta Thunberg's rise from a teen activist to a global ecological campaigner has inspired young people to follow her steps online and offline. Her world-famous campaign "Fridays For Future" followed by major hashtags #fridaysforfuture, #StayAtHome, #schoolstrike4climate, #ClimateStrike, #PeopleNotProfit has encouraged about 17 million people from more than 151 countries and 8,600 cities total (to join: *twitter.com/hashtag/FridaysForFuture?src=hashtag_click*). Digital technologies and social media enabled this involvement and set an unprecedented global activism example.

Moreover, this campaign is sustained mainly by social media and allows young people to remain connected despite distances. July 29, 2022, marked the 206th climate strike week. For this strike, some activists focused on their anniversaries and connected them to the current environmental problems. Sommer Ackerman, an activist from Finland, made a poster on Twitter (see Figure 8.2) about her 100 weeks of climate striking and CO_2 emission level going during this period *twitter.com/lifewithsommer/status/1552938411283152897*.

Sommer Ackerman
@lifewithsommer
29 July

Today marks my 100 weeks of climate striking with **#FridaysForFuture** and in that time atmospheric CO_2 has gone up by 1.512%. No more empty promises. Climate activists shouldn't have to exist in the first place.

366 Retweets   12 Quote Tweets   1,535 Likes

FIGURE 8.2   Sommer Ackerman's tweet

Other inspiring ecological campaigns include SustainUS with the most active representation on Facebook *facebook.com/hashtag/sustainus*, while Earth Guardians and Green Track are mostly featured on Twitter *cutt.ly/PVkUiqj*, *cutt.ly/CVkUGof*, etc. These initiatives effectively enable young uprising voices and reframe the global ecological framework, introducing the issues of empowerment, activism, responsibility, indigeneship, and justice.

In 2021 as a part of their course on Ecolinguistics, the students of Vasyl Stefanyk Precarpathian National University (Ukraine) were researching the ecological problems of their communities, analyzing global social media campaigns, and modeling the relevant social media campaigns for their communities. They were working with grassroots initiatives, researching local media, and connecting with community members to create a complete picture of their case. These social media campaigns presented ecological narratives for Facebook, Instagram, and Twitter. In addition, they exploited the affordances of social media to provide information and engage people in following, communicating on the topic online, and joining suggested events offline.

Most students come from the mid-sized towns of the Western region of Ukraine, situated partly in the Eastern Carpathian Foothills. They bring valuable local perspectives and implement decolonizing ecological initiatives. Decolonizing here becomes setting free from the dominating influence of superior organizations leading political and ecological agendas. Anthropocene has been previously viewed as explicitly linked with colonialism, even having its origin in colonial practices with disempowering social, economic, and ecological consequences (Davis & Todd, 2017).

Aiming to change the terms of the conversation about the ecological crisis, to reframe the narratives based on exclusion (Banerjee & Arjaliès, 2021), the focus has been mainly on the content and knowledge. In contrast, the producers of the reframed ecological discourse, especially in the education settings, should be equally salient. The social media affordances include strategic and discursive features that facilitate the engagement of the audience on social media through the interactive presentation of one's agenda. All the subsequent social media discourse examples come from these collaborative projects.

Strategic Features of Social Media Practices in the Ecolinguistics Course

The strategic features in this course include social media practices for best public outreach. Students focus on developing a coherent and systemic presence on social media with an awareness of their tools and structural applications.

Developing a Clear Voice

Making oneself heard on social media is a challenge that does not require perfection but strongly depends on consistency and individual input. The voice of young people may be heard only if it comes out consistently. Being present on social media and making the presence steady, not being overwhelming and off-putting followers is a prerequisite for social media success. Working on a specific project or promoting a distinct idea requires regularity and consistency. Once the new notion has entered the social media sphere, it needs to be constantly updated and kept within the scope of reach of interested parties. However, it also shows that focusing on quality, not quantity, serves better to achieve the goal. Also, developing one's voice is the key to building connections with the audience. Adding a personal note, a personal comment to everything reposted and posted online is a tool for deepening personal ties and creating a clear image of the social media user. It's important to be not solely a content generator but a person who has an idea to convey and uses social media for this.

Adopting a Systemic Approach

Every social media platform has specific features that can be used to make one's ideas visible and heard. Making individual social media posts using a personal page is the primary option for a social media user. This is the most effective way to make one's position clear and noticeable. However, posts can also be made on group pages, either specifically created for a specific project or any other relevant groups (local, professional, etc.). These groups may be private or public, influencing the users' involvement and feedback.

Once using a social media system, it is wise to conduct research concerning the previously published content, people, and organizations actively or passively involved in the topic. Others might as well use the systemic affordances they are using. Not only joining specific groups and spaces but also commenting on the pages of the major players in the field makes an impact. Social media nowadays allow static and dynamic content posting, so image/video sharing and stories/reels features enhance the audience reach.

Discursive Features of Social Media Practices in the Ecolinguistics Course

The discursive analysis of the social media reports and campaigns showed several linguistic milestones which became the architectural landmarks of the ecological social media discourse.

Hashtag Tactics

While social media users do not necessarily use headlines for their posts, which hinders a prompt recognition of the theme, it's important to allow them to navigate through the posts and trace the most recent updates. Ecological hashtags follow the pattern *(Noun(s)/Adjective/Verb/Pronoun) + Noun*, as in #eco #eco-projectangryearth #earth #planetearth #illustration #saveearth #saveour-planet #savetheearth #savetheworld #green 🌍 #ecofamilyworld.

Starting a new hashtag shouldn't be taken for granted. This is a content assignment, a creative and analytical way that can have a potential impact on the project's durability. In 2015, brand UCO started the eco-movement with the hashtag #TrashTag (or #trashtag), aiming to pick up 10,000 pieces of trash by 2016. This initiative got support from users on Instagram, Reddit, and Twitter over the weekend as they shared photos they took before and after and encouraged others to join (Nugent, 2019). Overall, as of August 2022, this hashtag has gathered more than 250,000 posts on Instagram only *instagram.com/explore/tags/trashtag*, and they are updated regularly, making this hashtag active for eight years.

Emoji Usage

Social media have brought new communication tools, ways of expressing oneself, and forms of literacy. Danesi (2020) sees emojis as intended for an informal register with structural, conceptual, and pragmatical features, allowing effective hybrid writing usage to emphasize the message's meaning or mark emotional rapport. The most widespread ecological symbols are emojis denoting objects of nature, human beings acting in an eco-friendly manner, artificial objects facilitating eco-friendly activities, global perspective emojis, and any other green emojis (hearts being the most popular) (see Figure 8.3). Emojis foster engagement on social media and make connections between social media content generators and users more emotional and effective.

Metaphors Usage

Metaphors are conceptual entities of dynamic character that "adapt to the discursive and practical needs of those who use them in socio-natural context"

🚶 Person walking	🌸 Cherry blossom	🌍 Globe showing Europe-Africa	🐞 Lady beetle
🚴 Person biking	🌺 Hibiscus	🌎 Globe showing Americas	🌲 Evergreen tree
🤸 Person cartwheeling	🌻 Sunflower	🌏 Globe showing Asia-Australia	🌴 Palm tree
🛴 Kick scooter	🌱 Seedling	🗺 World map	💚 Green heart

FIGURE 8.3   Popular Ecological Emojis

(Döring & Nerlich, 2005, p. 56). They manifest themselves non-linguistically in many aspects of contemporary life: building tall, levels of obesity, industrialization, use of time, traveling fast, urbanization, racial categorization and exclusion, medical practice, sexual behavior, militarization, evaluations of quality by quantity, the commodification of nature, treatment of animals, education and the concept of progress (Fill, 2010).

Metaphors influence and shape our worldviews. Environmental metaphors, apart from controlling, suppressing, misleading, liberating, and empowering, exemplify the role of metaphor in the perception of and interaction with the environment and are a source of replacement for metaphors that have disseminated and supported unsustainable way of life (Mühlhäusler, 2003, p. 140), or permanent destructive cognitive patterns in discourse (Stibbe, 2015, passim). Instead, the constructive and destructive metaphors create a strong impact and leave a lasting visual and verbal aftertaste.

Presenting the planet as a living being is one of the most powerful techniques to attract the attention of social media users. Human beings value life and recognize life/death relations instantaneously. While each metaphor highlights certain aspects of the concept and makes other aspects implicit, the metaphor X is A HUMAN BEING (X are PEOPLE) focuses on coming into existence, development, and going out of existence (G. Lakoff & Johnson, 1980, p. 201). This conceptual tool is inclusive concerning nature/planet Earth, as it creates a system of equal treatment traditionally related to human organizations and structures. Expanding the anthropocentric and anthropomorphic arrangements is timely, relevant, and easy to deconstruct.

The extreme danger to the planet may be exemplified by the intention to take away its life and stop its existence, as portrayed in Figure 8.4.

Metaphorical expressions may also summarize the post after the calls for action and aim to leave a lasting visual image with the readers (see Figure 8.5).

Thus, metaphors are a powerful tool for highlighting the features of nature and making eye-catchers for social media users.

FIGURE 8.4   Killing the Planet

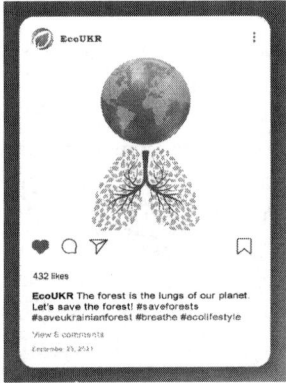

FIGURE 8.5   Saving the Planet

Oppositions

The role of oppositions or binarism/dichotomy/polarization/antonyms/contrast has been extensively studied and debated. (For an overview, see Lyons, 1977; Murphy, 2003.) Post-modernistic tendencies include multiple perspectives and the capability of the language to express the gradable and non-gradable opposites (Lyons, 1987; Davies, 2014). I use the term opposition to refer to any textual and contextual instance where discourse units are treated as opposing entities.

Contrasting ideas is one of the most effective devices aimed at representing two or more categories, highlighting their characteristics, and processing them as opposites on the conceptual level. For example, the most frequent oppositions in ecological social media discourse are gradable *global & local, small & big,* and *good & big* types.

Global and Local Opposition

Localization of the opposition in ecological, social media discourse sets the boundaries and views the local perspective in connection with the global scene. This may happen when the author mentions the place they are living in and emphasizes the necessity to remember the global perspective: *when thinking about the world's environmental problems, we must first take care of the cleanliness of the area where we live in.*

This example also prioritizes local over global issues, thus proving that the lenses of young people see the global problems through their habitat area. The interconnectedness of both perspectives is inherently present in this type of discourse. It leaves a lasting impression of the planet Earth being a kaleidoscope of "places" people are living in and should be taking care of: *in my opinion, the situation in my town is a miniature of the global situation to which humanity has led the planet.*

Young people should be free to use any historical examples they find relevant and recognizable to their peers. While some countries have a very limited set of examples identifiable worldwide, others are abundant in them. The example of Chornobyl is crucial for the Ukrainian youth. It has become a local and global warning to highlight the necessity of cooperation for the sake of preventing catastrophes: *in my opinion, the Ukrainian authority should ask for foreign partners' help, in order not to repeat the "Chornobyl" catastrophe.*

Means of people may also set distant perspectives/object nominations that are easily identifiable, respected, and able to become trendsetters:

- It's *not a shame to come to the office or the White House on a bicycle, on the contrary, this practice is appreciated and rewarded.*
- Shakespeare's *eternal question "To be or not to be" becomes "To sort or not to sort" in our town.*

Reframing the existing patterns boosts the visibility of the messages and, apart from communicating them visibly and clearly, enhances the multimodal potential of the social media post, allowing them to spark students' imagination.

Quality and Quantity Opposition

The issues of size and value are engraved into human consciousness and reflected in the grammar and semantics of languages. The ideology of growthism is profoundly entrenched in Western forms of speaking and is widely encountered also in so-called traditional societies (Mühlhäusler, 2003, p. 132). M.A. K. Halliday (2001) points out that the language system is focused on the gradeability of its properties with a negative and a positive pole and a correlation of good being big and small being bad (p. 194). In ecologically oriented discourses, the sustainable development discourse is the example of classical conceptual entities representing quality and quantity (*small is bad/big is good*) which coexist with novel units where small is treated as good (Goshylyk, 2017).

These conceptualizations aren't grounded in embodied experience but come from socially constructed awareness. The positive embodied experience proves that small things/entities grow and reach their optimal size. The example of children and grown-ups growing plants to their maximally useful state being fully grown provides a positive illustration of the quantity and quality relations (Mühlhäusler, 2003).

The complexity of the size and value distributions enables various discourse patterns to construct a vivid image of the required action. The positive relations between small actions are emphasized in social media posts with the sequence of smallness leading to greatness:

> I was born in Verkhovyna, Carpathian Mountains. We faced the problem of deforestation, the main environmental problem in our region, many years ago. There was a big flood in our area in 2008. Since our Carpathians are deforested, many people suffered a lot. There were landslides that left people homeless. The riverbanks broke off because there were no tree roots to hold them. My family also suffered, our land plot, not far from the forest, was split into two parts. My uncle was left without shelter and belongings. Right now, the residents of my town are involved in tree planting, teachers engage students, and district-wide tree planting days are happening regularly. We are trying to compensate for deforestation. Small steps to a larger goal.

Unlike the positive correlation between small and big, the negative one constructs a destructive image of possible consequences: *a small town in the Ivano-Frankivsk region, Burshtyn, poses a tremendous ecological threat to the whole country*. Regardless of the consequences of the actions described, the quantity of opposition is an effective tool for attracting attention.

Large quantities, which are traditionally viewed as positive, are generally reconsidered and reframed in ecological appeals. The following examples illustrate that young people want to reject consumerism (Figure 8.6) and prevent other artificial threats to nature and claim that reducing the number of products is an adequate solution (Figure 8.7).

Social media encourage sharing personal opinions and thus making exaggerations that can't be avoided. Qualitative and quantitative hyperboles create

FIGURE 8.6   Overconsumption Is a Threat

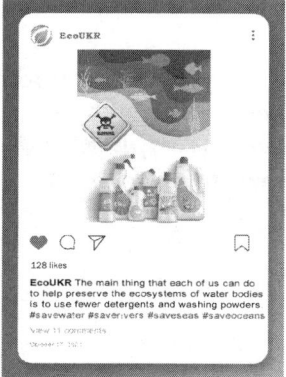

FIGURE 8.7   Less Is Good

strong images and make lasting impressions as in "*my hometown is one of the biggest environmental problems.*"

In social media posts, young people are also introducing positive descriptions and making overstatements afterward:

Ivano-Frankivsk is a beautiful city that for most of its citizens or tourists reminds of cozy and attractive European places. It is both modern and historical, progressive, and traditional. What isn't European about Ivano-Frankivsk is not caring about the pollution in the city and in the world.

This excerpt builds upon three nominations of *European places, the city of Ivano-Frankivsk,* and *the world. European* is used here in the meaning of characteristic of Europe and bears a positive connotation reinforced by adjectives *cozy, attractive, modern, historical, progressive*, and *traditional*. Other loci nominations form a hierarchy of responsibility for action and lack of actions on the local, continental, and global levels.

Personalization

Sharing personal experiences is one of the basic features of social media. In this type of media, people know that they are addressing their friends, relatives, acquaintances, and strangers. Regardless of the mixed type of addressees, personal experience is the starting point and the ultimate goal of lots of social media posts. Personalization online is the discursive strategy that shows the relation to and connection with a certain issue by constructing the image of a trustworthy social media user.

Personal experience serves several functions in environmental, social media discourse:

- **Young people set the example and call to action:** (1) *I want to change my village and ask you to join me!* (2) *Teach others to protect the environment, and do not be afraid to set an example. Everything is in our hands!*
- **Young people share their opinions:** *I truly believe that planting one tree can make a difference.* It's also important to teach students to differentiate between facts and opinions for sake of enhancing their critical thinking and media literacy.
- **Young people share their emotions:** (1) *I am very disappointed with the state of nature in my village.* (2) *I was frustrated when I saw this garbage all around. This can't happen again.*
- **Young people talk about intimate and family issues:** *I was born in Verkhovyna, Carpathian Mountains, and I can say that we faced deforestation many years ago. There was a big flood in our area in 2008 … My family also suffered … due to significant landslides, our land plot, not far from the forest, was split in half. My uncle was left without his house and belongings.*

Personalization is achieved by means of personal pronouns and sharing individual and group stories supported by relevant images. Pronouns are ultimately social in the speaker-listener shared reference and comprise the most frequently used function word category (Kacewicz et al., 2014). Pronouns are therefore treated as the signs of face-to-face conversation capable of establishing equality and intimacy with the audience. Personalization is best fitted with personal photos featuring the author of the post with, preferably, any piece of nature.

Inclusion

Role modeling in ecological social media posts may also relate to the group experience. This is consistent with the research claiming that the use of "we" correlates with more positive problem solutions within a relationship framework (Simmons et al., 2005). This involves considering that influencers reference the collective identity and use inclusive terms, such as "we," "us," and "our" in describing goals and achievements (Gardner & Avolio, 1998, p. 46). It also highlights that collective appeals are a way of producing a shared vision that justifies the frequency of inclusive phrases, pronouns in particular. These passages may include:

- **A call to action:** *At present, we can reduce CO emissions on our own: limit the burning of fuel at home, in electric generators, do not burn leaves, peatlands and dry land, do not smoke, and limit the use of vehicles.*
- **Empathizing inclusion:** *This issue requires public attention and important decision-making from local authorities. Together we can overcome this problem.*
- **Frequent repetition:** *WE are gradually but purposefully killing our planet. WE do not think about our future and our descendants who will live on clogged land. Every day, WE destroy ourselves by cutting down forests, throwing garbage in unassigned places, using disposable tableware and plastic bags, and other comfortable things for our daily lives.*

Future Research and Conclusions

Proactive ecological identity conceptualizes society-based identity construction in terms of informational awareness, performative orientation, and future risk mitigation. Social networks create fundamentally different opportunities in every aspect of these strands. For example, while ecological issues have always been on the agenda of governmental and political leaders, nonprofits, governmental agencies, and other major stakeholders, young people's concerns and visions weren't an eminent part of the dialogue. Occasionally, young people were given a voice, but social media allow them to take the initiative whenever needed, in whatever forms preferred, and with whatever messages required. Based on community knowledge and personal engagement, this performance becomes the key to decolonizing practices, thus enabling alternative visions of the future.

The involvement of young people, being the most loyal social media users, aims to alleviate ecological hazards in their communities and is facilitated using social media. While not taken for granted, young people's coherent and effective engagement on social media is facilitated by the knowledge of their structural features. This study identified the major types of strategic and discursive characteristics. It highlighted the ways of alleviating voices and utilizing systemic patterns, as well as hashtags, metaphors, emojis, oppositions, personalization, and inclusion applying. A reverse information flow model documents the use of understanding, evaluating, and generating content. A critical analysis of discourse contributes to our understanding of the architecture of youth activism presented online and provides a basis for its educational elaboration.

Though this study was limited by the types of social media and the number of languages that have been analyzed, leaving out useful tools and philosophies, it has combined productive research approaches to an ecological discourse and set a coherent eco-educational paradigm. Young people, in collaboration with educators and social researchers of various backgrounds, can expand the horizons of decolonizing attempts, deconstructing, and reconstructing ecological sovereignty.

Acknowledgment

The work on the course "Ecolinguistics" at Vasyl Stefanyk Precarpathian University (Ukraine) was partially funded by the European Union in the framework of the Foreign Language Teacher Training Capacity Development as a Way to Ukraine's Multilingual Education and European Integration – MultiEd project (Project Number: 610427-EPP-1-2019-1-EE-EPPKA2-CBHE-JP) under the ERASMUS+ program. This document does not represent the opinion of the European Union, and the European Union is not responsible for any use that might be made of its content.

References

Albrechtslund, A. (2013). New media and changing perceptions of surveillance. In J. Hartley, J. Burgess, & A. Bruns (Eds.), *A companion to new media dynamics* (pp. 309–321). Wiley Blackwell. doi:10.1002/9781118321607.ch19

Banerjee, S. B., & Arjaliès, D.-L. (2021). Celebrating the end of enlightenment: Organization theory in the age of the Anthropocene and Gaia (and why neither is the solution to our ecological crisis). *Organization Theory, 2*(4). doi:10.1177/26317877211036714

Bauman, R. (2000). Language, identity, performance. *Pragmatics: Quarterly Publication of the International Pragmatics Association, 10*(1), 1–5. doi:10.1075/prag.10.1.01bau

Bauman, R., & Briggs, C. (1990). Poetics and performance as critical perspectives on language and social life. *Annual Review of Anthropology, 19*, 59–88. doi:10.1146/annurev.anthro.19.1.59

Beach, R., & Smith, B. E. (2020). Using digital tools for studying about and addressing climate change. In P. M. Sullivan, J. L. Lantz, & B. A. Sullivan (Eds.), *Handbook of research on integrating digital technology with literacy pedagogies* (pp. 346–370). IGI Global.

Berger, P. L., & Luckmann, T. (1991). *The social construction of reality.* Penguin.

Bergmann, C., Dimitrova, N., & Alaslani, K., et al. (2022). Young children's screen time during the first COVID-19 lockdown in 12 countries. *Scientific Reports, 12,* 2015. doi:10.1038/s41598-022-05840-5

Biden, J. (2019). *The Biden plan for a clean energy revolution and environmental justice.* https://joebiden.com/climate-plan/

Bloom, B. S., & Anderson, L. W. (2014). *A taxonomy for learning, teaching, and assessing: A revision of Bloom's.* Pearson.

Bloomfield, E. F., & Tillery, D. (2019). The circulation of climate change denial online: Rhetorical and networking strategies on Facebook. *Environmental Communication, 13*(1), 23–34. doi:10.1080/17524032.2018.1527378

Boulianne, M., Lalancette, M., & Ilkiw, D. (2020). "School Strike 4 climate": Social media and the international youth protest on climate change. *Media and Communication (Lisboa), 8*(2S1), 208–218. doi:10.17645/mac.v8i2.2768

Boulianne, M., & Theocharis, Y. (2020). Young people, digital media, and engagement: A meta-analysis of research. *Social Science Computer Review, 38*(2), 111–127. doi:10.1177/0894439318814190

Butler, J. (1990/1999). *Gender trouble: Feminism and the subversion of identity.* Routledge.

Cameron, S., Russell, A., Brake, L., Fredlund, K., & Morris, A. (2022). Twitter users' displays of affect in the global warming debate. *Journal of Technical Writing and Communication, 52*(2), 213–235. doi:10.1177/00472816211007804

Castree, N. (2021). Framing, deframing and reframing the Anthropocene: This article belongs to Ambio's 50th Anniversary Collection. Theme: Anthropocene. *Ambio, 50*(10), 1788–1792. doi:10.1007/s13280-020-01437-2

Christoff, P., & Eckersley, R. (2013). *Globalization and the environment.* Rowman & Littlefield.

Comfort, S. E., & Hester, J. B. (2019). Three dimensions of social media messaging success by environmental NGOs. *Environmental Communication, 13*(3), 281–286. doi: 10.1080/17524032.2019.1579746

Damico, J. S., Baildon, M., & Panos, A. (2020). Climate justice literacy: Stories we live by, ecolinguistics, and classroom practice. *Journal of Adolescent & Adult Literacy, 63*(6), 683–691. doi:10.1002/jaal.1051

Danesi, M. (2020). *The semiotics of emoji: The rise of visual language in the age of the internet.* Bloomsbury.

Davies, M. (2014). *Oppositions and ideology in news discourse.* Bloomsbury.

Davis, H., & Todd, Z. (2017). On the importance of a date, or, decolonizing the Anthropocene. *ACME: An International Journal for Critical Geographies, 16,* 761–780.

De Fina, A. (2006). Group identity, narrative and self-representations. In A. De Fina, D. Schiffrin, & M. G. Bamberg (Eds.), *Discourse and identity* (pp. 351–375). Cambridge University Press.

De Fina, A., Schiffrin, D., & Bamberg, M. G. W. (2006). *Discourse and identity.* Cambridge University Press.

Deignan, A., & Semino, E. (2022). Metaphors and meaning-making in young people's talk about climate change. In H. L. Colston, T. Matlock, & and G. J. Steen (Eds.), *Dynamism in metaphor and beyond. Metaphor in language, cognition and communication* (pp. 191–204). John Benjamins.

Döring, M., & Nerlich, B. (2005). Assessing the topology of semantic change: From linguistic fields to ecolinguistics. *Logos and Language: Journal of General Linguistics and Language Theory, 6*(1), 55–68.

Entman, R., & Usher, N. (2018). Framing in a fractured democracy: Impacts of digital technology on ideology, power and cascading network activation. *Communication Theory, 68*(2), 298–308. doi:10.1093/joc/jqx019

European Commission. (2019). *Communication from the Commission to the European Parliament, the European Council, the Council, the European Economic and Social Committee and the Committee of the Regions—The European Green Deal (COM (2019) 640 final, 11.12.2019).* https://eur-lex.europa.eu/legal-content/EN/TXT/?uri=LEGISSUM:4438420

European Parliament & Council of the European Union. (2022). *Decision (EU) 2022/591 of the European Parliament and of the Council of 6 April 2022 on a General Union Environment Action Programme to 2030.* https://eur-lex.europa.eu/legal-content/EN/TXT/?uri=CELEX:32022D0591

Field, E. (2021). Is it all just emojis and LOL: Or can social media foster environmental learning and activism? In M. Hoechsmann, G. Thésée, & P. R. Carr (Eds.), *Education for democracy 2.0* (pp. 198–220). Brill.

Fill, A. (2010). *The language impact. Evolution - system - discourse.* Equinox.

Gardner, W., & Avolio, B. (1998). The charismatic relationship: A dramaturgical perspective. *The Academy of Management Review, 23*(1), 32–58. doi:10.2307/259098

Goshylyk, N. (2017). "Small is beautiful" in English mass media texts on sustainable development. *Arbeiten Aus Anglistik Und Amerikanistik, 42*(1), 141–158.

Halliday, M. A. K. (2001). New ways of meaning: The challenge to applied linguistics. In A. Fill & P. Mühlhäusler (Eds.), *The ecolinguistics reader: Language, ecology, and environment* (pp. 175–202). Routledge.

Hartley, J., Burgess, J. E., & Bruns, A. (Eds.). (2013). *A companion to new media dynamics.* Wiley-Blackwell.

Hoffman, A. J., & Jennings, P. D. (2018). *Re-engaging with sustainability in the Anthropocene era: An institutional approach.* Cambridge University Press.

Howe, C. (2019). *Ecologics: Wind and power in the Anthropocene.* Duke University Press.

Kacewicz, E., Pennebaker, J. W., Davis, M., Jeon, M., & Graesser, A. C. (2014). Pronoun use reflects standings in social hierarchies. *Journal of Language and Social Psychology, 33*(2), 125–143. doi:10.1177/0261927X13502654

Krathwohl, D. R. (2002). A revision of Bloom's taxonomy: An overview. *Theory into Practice, 41*(4), 212–218. doi:10.1207/s15430421tip4104_2

Lakoff, G., & Johnson, M. (1980). *Metaphors we live by.* University of Chicago Press.

Lakoff, R. T. (2006). Identity à la carte: You are what you eat. In A. De Fina, D. Schiffrin, & M. G. Bamberg (Eds.), *Discourse and identity* (pp. 142–165). Cambridge University Press.

Landis, B. T., & Allen, W. L. (2022). Cascading activation revisited: How audiences contribute to news agendas using social media. *Digital Journalism, 10*(4), 537–555. doi:10.1080/21670811.2021.1962728

Lee, N. M., VanDyke, M. S., & Cummins, R. G. (2018). A missed opportunity?: NOAA's use of social media to communicate climate science. *Environmental Communication, 12*(2), 274–283. doi:10.1080/17524032.2016.1269825

Lewis, S. L., & Maslin, M. (2018). *The human planet: How we created the Anthropocene.* Yale University Press.

Liu, B. F., & Kim, J. (2022). Social media and climate change dialogue: A review of the research and guidance for science communicators. In J.-W. Yusuf & B. St. John (Eds.), *Communicating climate change: Making environmental messaging accessible* (pp. 97–115). Routledge.

Loukianov, A., Burningham, K., & Jackson, T. (2020). Young people, good life narratives, and sustainable futures: The case of Instagram. *Sustainable Earth, 3*(1), 1–14. doi:10.1186/s42055-020-00033-2

Lyons, J. (1977). *Semantics: John Lyons.* Cambridge University Press.

Lyons, J. (1987). *New horizons in linguistics 2.* Penguin.

Mühlhäusler, P. (2003). *Language of environment, environment of language: A course in ecolinguistics.* Battlebridge.

Murphy, M. L. (2003). *Semantic relations and the lexicon: Antonymy, synonymy and other paradigms.* Cambridge University Press.

Napawan, N. C., Simpson, S. A., & Snyder, B. (2017). Engaging youth in climate resilience planning with social media: Lessons from #OurChangingClimate. *Urban Planning, 2*(4), 51–63. doi:10.17645/up.v2i4.1010

Nugent, C. (2019). People are picking up trash in parks and beaches for the "Trashtag Challenge." [Weblog post]. https://time.com/5548966/trashtag/

Page, R. M., & Page, T. S. (2000). *Fostering emotional well-being in the classroom* (2nd ed.). Jones and Bartlett.

Parker, S. K., Bindl, U. K., & Strauss, K. (2010). Making things happen: A model of proactive motivation. *Journal of Management, 36*(4), 827–856. doi:10.1177/0149206310363732

Simmons, R. A., Gordon, P. C., & Chambless, D. L. (2005). Pronouns in marital interaction: What do "you" and "I" say about marital health? *Psychological Science, 16*(12), 932–936. doi:10.1111/j.1467-9280.2005.01639.x

Singh, R. L., & Singh, P. K. (2016). Global environmental problems. In R. L. Singh (Ed.), *Principles and applications of environmental biotechnology for a sustainable future* (pp. 13–41). Springer.

Sklair, L. (Ed.). (2021). *The anthropocene in global media: Neutralizing the risk.* Routledge.

Smith, B., Beach, R., & Shen, J. (2021). Fostering student activism about the climate crisis through digital multimodal narratives. *The Journal of Sustainability Education,* 25. http://www.susted.com/wordpress/content/fostering-student-activism-about-the-climate-crisis-through-digital-multimodal-narratives_2021_08/

Stibbe, A. (2015). *Ecolinguistics: Language, ecology and the stories we live by.* Routledge.

Thomashow, M. (1995). *Ecological identity: Becoming a reflective environmentalist.* MIT Press.

Tyson, A., Kennedy, B., & Funk, C. (2021). Gen Z, millennials stand out for climate change activism, social media engagement with issue. Pew Research Center. https://www.proquest.com/reports/gen-z-millennials-stand-out-climate-change/docview/2552826935/se-2

United Nations. (2015). *The 2030 agenda for sustainable development.* https://sdgs.un.org/publications/transforming-our-world-2030-agenda-sustainable-development-17981

SECTION III

Providing Students with Media Production Methods to Enact Change

Students are more likely to engage in media production when their teachers or adult leaders provide support and instructions on effective media production. The final section of this book describes the need for teachers and adult leaders in organizations to provide students with this support and instructions on producing media on climate change (Bernier, 2020; Boardman et al., 2021).

Teachers and adult leaders can provide this support for students by having them engage in the following practices:

- *Articulate their purpose for producing media* related to portraying and documenting climate change impacts for achieving change through positive audience uptake (Leal Filho & Pace, 2016; Martusewicz et al., 2015; Robertson & Westerman, 2015; Turner, 2015). Producing media for audiences requires fostering change in their knowledge and attitudes, leading to action (Beach, 2022; Beach & Smith, 2020). For example, students may use videos to portray climate change impacts such as flooding, droughts, or sea rise to motivate audiences to take action to address these impacts (Franzen & Mader, 2020; Lee et al., 2020).
- *Draw on students' digital media experiences* to help them transfer those experiences to their media production. Nineteen percent of adolescents indicate that they use YouTube; 16% use TikTok; and 15% use Snapchat almost constantly (Vogels et al., 2022). They can draw on their use of these platforms to produce their media for visually replicating audiences' experiences with climate change impacts (Isik & Vessel, 2021).
- *Draw on students' discussions* of the climate crisis with peers or family members that strongly predict students' climate change activism, particularly through their social media interactions (Valdez et al., 2018).

DOI: 10.4324/9781003335276-12

- *Employ writing to envision and plan for* multimodal media production, for example, writing scripts or storyboards for video or podcast productions as well as generating narratives, reports, and poetry about climate change (Beach, 2015; Dobrin, 2015; Smith, B. E., 2017, 2019).
- *Engage with environmental issues and spaces* through hands-on activities. For example, engaging in field trips to study climate change impacts to then draw on those experiences to imagine alternative futures through the use of adaptation and mitigation practices for addressing climate change (Engle-Friedman et al., 2022). Preservice teachers as students in The Rivers2Lake Education Program *t.ly/Pubw* in Northern Wisconsin interact with members of the Anishinaabe (Ojibwe) people based on hands-on interactions with rivers and Lake Superior to study climate change impacts on rivers and lakes (Ernst et al., 2020).
- *Provide students with options* for producing different types of media—video, digital stories, podcasts, blog posts, social media, artwork, music, art, narratives, poetry, written reports, etc. Students may select which media they believe will best communicate their ideas and perspectives to audiences beyond their classrooms. For example, students may use blogs (for a list of blogs *t.ly/tKIE*) or sites for sharing media *t.ly/la-B*, podcasts *t.ly/b3AH*, or writing *t.ly/_ri2*, for example, writing for Youth Voices for Climate Action *t.ly/VRGB*.
- *Provide training* for students in specific media production methods, for example, techniques for creating and editing videos using different video production platforms (Bryne, 2022, *t.ly/NKtW*). For example, students in the previously mentioned Lens on Climate Change (LOCC) project at the University of Colorado (Littrell et al., 2020, 2022), receive extensive instruction on how to create scripts and storyboards, engage in interviews, employ video cameras for the use of different types of shots and angles, and edit their videos (for a description of this instruction, see *t.ly/WVro*).

  Here are three videos from the LOCC project:
  - Streamflow on the Crystal (the challenge of a local Colorado river drying up due to drought) *youtube.com/watch?v=Layk4xcio-8*
  - Wildfires in the West (the problem of increasing drought in Colorado leading to more wildfires) *youtube.com/watch?v=_MdiiDF-f88&t=1s*
  - Behind the Panels (the value of using solar panels as alternative energy) *youtube.com/watch?v=bGcYlS_EwJY&t=4s*

    Students also need to know how to submit videos to competitive projects based on criteria for judging their entries to also learn how to apply these criteria for improving their videos as well as receiving recognition for their work, for example, the Videos for Change project *videosforchange. org/global-showcase*. One example of a winning video in the EmPower Silicon Valley Contest: Sustainable Transportation Builds Community (portrays the need to transition to sustainable transportation) *youtube.com/ watch?v=LMSVdJbOOa4*

- *Create organizations* within schools, for example, environmental clubs, that support students to work collaboratively on projects to engage their peers in sustainability practices (Smith, W., 2019). For example, students in environmental clubs can create school gardens for understanding regenerative agriculture practices or for promoting healthy food options in the school cafeteria. Students in these organizations can also produce media about climate change for sharing with peers, teachers, and administrators, for example, posters to put up in school hallways or writing for the school newspaper/website (for examples of school-related organizations *t.ly/lxJ1* or national organizations supporting school-related organizations *t.ly/iX24*).
- *Assess students' production* by developing criteria for assessing students' media production related to the potential or actual uptake from audiences. These criteria may include students' formulation or use of:
  - *A clear sense of purpose* related to the uptake they are attempting to achieve, for example, to convince audiences of the adverse impacts of plastics in the ocean or to imply the need for audiences to take actions to address a certain impact.
  - *Current, valid information and ideas* on climate change impacts and solutions at both the local and global levels.
  - *Visual images* in videos, digital stories, artwork, or posters, as well as the use of *cinematic techniques and editing*, that serve to engage audiences related to portraying certain climate change impacts.
  - *Audio or voice-overs* in videos, digital stories, podcasts, or music productions for engaging audiences.
  - *Interviews with or quotes* from people/policymakers for videos, digital stories, podcasts, social media, or written reports related to their voicing relevant, insightful perspectives on climate change.
  - *Engaging, readable writing, scripts, or storyboards* about climate change or for preparing students for media production.

    Teachers' ability to assess students in these instructional activities may vary according to their knowledge, experience, interest, and attitudes regarding climate change.
- Given that preservice or beginning teachers may have background experiences in supporting students' media production, school districts and/or teacher education programs need to provide teachers with training and professional development on media production practices related to teaching about climate change. Central to such training is having teachers engage in their own media production practices to then model their experiences in media production for their students.

Chapter Summaries

The first chapter in this section, "Climate Writing Across Media: Scribing New Stories-to-Live-By" by Antonio López, describes specific ecoliteracy practices for planning various types of media production projects based on thinking about audiences. He addresses the need for teachers to instruct students on how to employ writing to envision, propose, and plan for their productions as an alternative to actual media production. For example, students create scripts and storyboards for video or podcast productions, as illustrated by case study examples from López's undergraduate college courses applied to blogging, video essays, and digital storytelling for climate communication. Then, students formulated a proposal and a press kit for disseminating media to larger audiences to identify a particular communication style relevant to audience uptake.

The second chapter, "The Long Haul: Three Decades of Teaching Student Documentary Action Research for Environmental and Climate Justice" by Steve Goodman, describes specific examples of training youth in after-school workshops at the Educational Video Center (EVC) in New York City. In this project, students create and edit video documentaries to portray local climate change effects using critical inquiry/project-based strategies and youth participatory action research. The chapter cites specific examples of interviews with former EVC teachers and students spanning three decades describing video production about students' portrayals of climate change impacts on local neighborhoods, schools, and homes. Students also engaged in participatory action research about environmental justice issues, creating videos portraying their work as climate change activists in their communities.

The following chapter, "Resilient by Youth Engagement: The Alameda Creek Atlas" by N. Claire Napawan, Brett L. Snyder, and Beth Ferguson, describes a place-based project, the *Alameda Creek Atlas*, involving youth employing social media and art/mapping for addressing sustainability issues in a local creek that was the selected finalist team, Public Sediment, for the 2017–2018 Rockefeller funded Resilient by Design, Bay Area Challenge. The chapter describes how students learned to employ various tools for collecting data about climate change impacts on fish as well as water quality. In addition, they employed social media and images for observing, documenting, and communicating concerns about these impacts and fostering organized events related to taking actions to address these impacts, examples with implications for teachers' use of social media and images.

The final chapter in the book, "Elevating Young Voices: Media Created by Youth for Youth" by Julian Arenas, Ardra Charath, and Liane Xu, describes how they, as young people, engage in editing The Youth Think Climate (YTC) magazine, a digital magazine with the support of Action for the Climate Emergency that includes essays, narratives, poetry, art, and music by and for young people throughout the globe. The magazine, therefore, provides a space for youth to express their perceptions, emotions, anxieties, and beliefs to address themes of climate denial and justice. The authors reflect on their practices as editors of YTC on how and why they select and support submissions based on the need to engage their youth audiences, enhancing those audiences to generate their media within a range of different contexts. Given that this book focuses on youth-created media, we were pleased that we had young people themselves writing this chapter to give them the final word on this topic.

For additional links, resources, and readings on the book's website related to this section, refer *t.ly/blEiD*.

References

Beach, R. (2015). Imagining a future for the planet through literature, writing, images, and drama. *Journal of Adolescent & Adult Literacy, 59*(1), 7–13.

Beach, R. (2022, February 15). A conversation with Richard Beach [Podcast]. Classroom Caffeine. https://www.classroomcaffeine.com/guests/richard-beach

Beach, R., & Smith, B. E. (2020). Using digital tools for studying about and addressing climate change. In P.M. Sullivan, J.L. Lantz, & B.A. Sullivan (Eds.), *Handbook of research on integrating digital technology with literacy pedagogies* (pp. 346–370). IGI Global.

Bernier, A. (2020). Wanting to share: How integration of digital media literacy supports student participatory culture in 21st-century sustainability education. *Journal of Sustainability Education, 24.* ISSN: 2151-7452.

Boardman, A. G., Garcia, A., Dalton, B., & Polman, J. L. (2021). *Compose our world: Project-based learning in secondary English language arts.* Teachers College Press.

Bryne, R. (2022, September 22). *My updated big list of tools for a variety of classroom video projects* [Web log post]. http://t.ly/NKtW

Dobrin, S. I. (Ed.). (2015). *Ecology, writing theory, and new media writing ecology.* Routledge.

Engle-Friedman, M., Tipaldo, J., Piskorski, N., Young, S. G., & Rong, C. (2022). Enhancing environmental resource sustainability by imagining oneself in the future. *Journal of Environmental Psychology, 79,* 101746.

Ernst, J., Erickson, D., Burgess, E., & Feldbrugge, R. (2020, December). Beyond traditional teacher professional development: Innovations in teacher professional learning in environmental and sustainability education. *Journal of Sustainability Education.* http://t.ly/W_oQ

Franzen, A., & Mader, S. (2020). Can climate skeptics be convinced? The effect of nature videos on environmental concern. *Sustainability, 12*(7), 2972. doi:10.3390/su12072972

Isik, A. I., & Vessel, E. A. (2021). From visual perception to aesthetic appeal: Brain responses to aesthetically appealing natural landscape movies. *Frontiers in Human Neuroscience. 15,* 676032. doi:10.3389/fnhum.2021.676032

Leal Filho, W., & Pace, P. (2016). *Teaching education for sustainable development at university level.* Springer.

Lee, K., Gjersoe, N., O'Neill, S., & Barnett, J. (2020). Youth perceptions of climate change: A narrative synthesis. *Wiley Interdisciplinary Review of Climate Change, 11*(3), e641.

Littrell, M. K., Gold, A. U., Koskey, K. L. K., May, T. A., Leckey, E., & Okochi, C. (2022). Transformative experience in an informal science learning program about climate change. *Journal of Research on Science Teaching*, 1–25. doi:10.1002/tea.21750

Littrell, M. K., Tayne, K., Okochi, C., Leckey, E., Gold, A. U., & Lynds, S. (2020). Student perspectives on climate change through place-based filmmaking. *Environmental Education Research*, *26*(4), 594–610. doi:10.1080/13504622.2020.1736516

Martusewicz, R. A., Edmundson, J., & Lupinacci, J. (2015). *Ecojustice education: Toward diverse, democratic, and sustainable communities*. Routledge.

Robertson, D., & Westerman, J. (Eds.). (2015). *Working on earth: Class and environmental justice*. University of Nevada Press.

Smith, B. E. (2017). Composing across modes: A comparative analysis of adolescents' multimodal composing processes. *Learning, Media & Technology*, *42*(3), 259–278.

Smith, B. E. (2019). Mediational modalities: Adolescents collaboratively interpreting literature through digital multimodal composing. *Research in the Teaching of English*, *53*(3), 197–222.

Smith, W. (2019). The role of environment clubs in promoting ecocentrism in secondary schools: Student identity and relationship to the earth. *The Journal of Environmental Education*, *50*(1), 52–71. doi:10.1080/00958964.2018.1499603

Turner, R. J. (2015). *Teaching for ecojustice*. Routledge.

Valdez, R. X., Peterson, M. N., & Stevenson, K. T. (2018). How communication with teachers, family and friends contributes to predicting climate change behaviour among adolescents. *Environmental Conservation*, *45*(2), 183–191. doi:10.1017/S0376892917000443

Vogels, E. A., Gelles-Watnick, R., & Massarat, D. (2022, August 10). *Teens, social media and technology 2022*. Pew Research Center. https://www.pewresearch.org/internet/2022/08/10/teens-social-media-and-technology-2022/

9

CLIMATE WRITING ACROSS MEDIA

Scribing New Stories-to-Live-By

Antonio López

Whether it's video, podcasting, blogging, comics, social media, or digital storytelling, all media projects require a blueprint for manifestation. To prepare students to thrive in the media world, an essential skill they should develop is writing the most important documents that bookend any media project: proposals and media kits. The proposal is the seed from which the media project will grow and receive nutrients. And for the media project to succeed and find an audience, it must be harvested into a well-crafted media kit to launch its presence in the world.

This chapter shares a methodology for proposal writing and media kits that was developed from teaching core communications curriculum courses to undergraduates. As a result of the COVID-19 pandemic and the switch to remote and hybrid teaching, proposal writing has become an effective pedagogical alternative to actual media production, which is prohibitive to students without access to high bandwidth or technology. This chapter draws from case studies to identify core practices for media projects that can be applied to climate writing. These practices include refining a communication style for targeted audiences, grounding messaging in values, clarifying motives, understanding medium properties, and identifying venues for distribution.

Climate advocacy media starts with writing. This comes across as counterintuitive to most students that are accustomed to quickly producing and uploading media into social networks with easy plug-and-play apps. However, writing is the magic key for opening doors to publishing, funding, and green lighting projects, and, most importantly, it's how ideas are test-driven without any cost of failure. Writing is essential for researching, planning, designing, and scripting any climate communication regardless of the media utilized—memes,

DOI: 10.4324/9781003335276-13

editorials, journalism, public service announcements (PSAs), Instagram posts, or TikTok videos. No matter how "new" the technology or platform, old-school writing remains the core of media planning and execution.

As the ecological crisis becomes more acute, media writing skills urgently need to be deployed in the service of climate stories. The dominant myths of our culture are stories-we-live-by—"stories in the minds of multiple individuals across cultures" (Stibbe, 2015, p. 6). These stories offer grand narratives that we live within and are, therefore, difficult to recognize (in the same sense, how ideology is a taken-for-granted belief system).

The "story" we see reproduced through the media and in all our mainstream institutions (including education) is that technological progress benefits all humans and unlimited economic growth is beneficial. This serves as a mental model for how we engage the world and how we should act with or upon living systems. The climate crisis is the product of a story founded on human exceptionalism: Earth exists merely for human comfort; we are free to consume or destroy natural "resources" at will but are safe from destruction ourselves; the extinction or eradication of other species does not matter; and the Earth will continue to sustain us even if we do not sustain the Earth. The power of the "old" story is based on fossil fuel culture and students need to clarify what they want to change about it (Corbett, 2021).

Ecolinguistics, the study of how language patterns our worldview, offers not only a critique of existing climate stories but also maps for generating new ones. Stibbe (2015) identifies three kinds of ecological stories: beneficial, destructive, and ambivalent. A beneficial story promotes climate action, whereas a destructive discourse harms or leads to climate inaction (as is the case with fossil fuel industry disinformation). Ambivalent stories are those with outcomes that are not clear; they may convey positive messages about the environment but might also serve as a form of harmful greenwashing. As is the case with the Anthropocene story—which blames all humans equally for the climate crisis— it's important to recognize that

> The 'we' in stories-we-live-by is not to suggest that all people in a society unwittingly acquiesce to these stories in the same way. Some individuals, cultural groups, and communities resist or defy these stories, and some have been doing it for decades or centuries. Stories-we-live-by are not universal, coursing through all cultures in the same way.
>
> *Damico and Baildon (2022, p. 14)*

For example, climate youth activists from Pacific Island Countries and Africa are engaging stories that counter dominant myths perpetuated by the global economy and international institutions.

In *Writing for Advocacy: The Climate Crisis*, I teach writing across different media to undergraduate students. This instruction is based on the assertion that writers and media creators need to recognize how environmental beliefs are tied to identity and values to communicate about the climate and environment effectively. An effective environmental communicator requires targeting and developing appropriate communication strategies for intended audiences that tap into their values.

In this chapter, I walk through a series of activities and assignments that I use with students. I start with writing activities to explore personal values. Next, I describe different digital writing projects that build up skills to arrive at the core writing skills students should learn for proposal writing and media kits. As we educate and train future climate communicators, proposal writing is one of the most fundamental skills students need to master. Another crucial tool is the media kit (usually comprising a cover letter and press release) that gets projects publicized and gains traction in the media world. The written proposal and media kit are bookends for any successful media venture. However, first, I offer a personal context for how this methodology was developed.

From "Old School" to "New School": Lessons from Working in the Media World

My teaching approach is inspired by my experience in the Los Angeles punk scene during the 1980s when I co-published a local music zine (a small, hand-made fan publication) with my neighborhood friends. Creating a zine as a teenager taught me the value of having a voice and a sense of empowerment from self-publishing, an early example of a participatory culture that existed before the interactive web. Learning publishing and journalism was largely self-taught. While there was certainly peer mentoring in the scene, we were amateurs and primarily learned by doing, which was more do-it-with-others (DIWO) than do-it-yourself (DIY). This kind of experiential learning is something I emphasize in my own pedagogy, so I always make sure that students make something.

As my interests grew beyond zine production to professionalization, I learned on the fly how to write grants and propose stories to editors. I quickly understood that the only way to get paid or advance in journalism and publishing was to master the query letter, a kind of elevator pitch to editors that determines whether a writing project leads to a contract. As a result, I probably spent more time penning query letters than writing for publication. Preparing the query meant researching the publisher and editor, preparing the background materials to justify the article, determining the right tone for the letter, and writing precisely and clearly.

Developing these skills enabled me to work professionally as a journalist for many years, which led to another key insight I convey to students. After being a freelancer for a local daily newspaper in northern New Mexico, I was offered a regular job to write for its weekend arts and entertainment supplement. The paper served a small city of 70,000 and was an important clearinghouse of news for the arts community. Despite doing my best to keep a beat on what was happening, inevitably, I would get an angry phone call or email on Friday morning, the day of publication. Some irate individual would demand an explanation about why their event or activity had not been written up in the paper. I would explain that working journalists and editors are busy with early deadlines and rely on outreach efforts to keep abreast of what is happening in the community.

These occasional calls also made clear that artists, galleries, museums, promoters, venues, and organizations that were better resourced and knew how to "work" the media got covered; amateur or less experienced artists and organizations lacked the skills to do media outreach. Professional media kits and press releases, the "currency" of media, reduce the workload of media decision-makers, so when utilized effectively, they are more likely to get attention and coverage. This led me to develop a workshop, "The Press Isn't Psychic," to train less-resourced communities on communicating with the media to get the coverage they deserve. Likewise, educators who teach media projects may not have experience working in the media trenches, so it is unlikely that outreach targeted at the media is incorporated into assignments. Thus, I determined that educators should learn these skills too.

Upon entering academia, I committed to developing for my department a core undergraduate course, *Writing Across the Media*, to ensure that proposal and promotion skills were a part of the required curriculum. Baked into the structure of *Writing Across the Media* is an emphasis on seven core communication principles that apply to any media project:

1 Clarify your issue, core values, and purpose.
2 Define your goal by identifying your desired outcomes.
3 Know who you want to reach and determine your audience.
4 Strategize the correct language and tone to reach that audience.
5 Specify the appropriate platform and medium and write for that medium.
6 Develop a distribution and outreach strategy.
7 Have a plan to follow up and track success and failure.

This chapter focuses on steps 1–5 (to use these assignments for information literacy, see López et al., 2020).

It's also important to emphasize the core ingredients of memorable storytelling, best summarized in the Simple Unexpected Concrete Credible Emotional

Stories (SUCCES) model popularized in the book, *Made to Stick* (Heath & Heath, 2008). First, *simplify* to the point that a clear take-home message can be summarized in one sentence. *Unexpected* means surprising the audience by inviting them to learn something they didn't already know or enabling them to see something different from conventional wisdom. People relate better to a story if there are *concrete* details that connect to their lived experience; effective storytelling ties the imagination to sensory experience. This should be aligned with empirical evidence that makes the story *credible*. Finally, audiences will have an *emotional* response when the story connects to values and core motivations for actions and beliefs. All these elements combine into *stories* that simulate and trigger the imagination to act as a theater of the mind. Every project in this chapter benefits from these key insights.

Many students have told me that the skills they learned in *Writing Across the Media* were essential for their success in the professional world. Their entry-level jobs (often as communications officers) almost always require social media posts, writing press releases, and performing outreach. Likewise, these skills are essential for environmental and climate communication.

Clarifying Climate Values

Before embarking on projects, beginning media production students are often asked to clarify the three "whys": Why this topic? Why now? Why are you the one to do it? This section offers an overview of how to answer these essential questions for climate storytelling.

In recent years, the concept of purpose-driven marketing has emerged, which orients communication goals with social causes and core values. This is applied to selling products (BP: "We care deeply about how we deliver energy to the world") or activism ("Greenpeace's goal is to ensure the ability of Earth to nurture life in all its diversity"). Purpose-driven marketing is based on three levels:

1 Determining what is important that demands action and drawing attention to (the *issue*).
2 Establishing the reason and motive for doing something (*purpose*).
3 Recognizing what is important and the principles at stake (*values*).

In the case of climate communication, Corbett (2021) notes that change is not science down but values up. Typically, people are not rationally motivated by scientific evidence, so there needs to be an appeal to their values and emotions. This is also confirmed by other researchers who argue that the "banking" or information deficit model of science and information literacy usually doesn't work (Damico & Baildon, 2022; Kellner & Share, 2019).

If students want to reach an audience with a different framework from theirs, they need to understand that people have diverse assumptions about human behavior and the goals of the sociopolitical system; these beliefs tend to align with political views. Not surprisingly, climate deniers tend to have right-wing political views, whereas those who acknowledge that the climate crisis demands action are usually on the opposite end of the political spectrum. For example, in the US, research shows that there are six "global warming Americas" with "different psychological, cultural, and political reasons for acting—or not acting—to reduce greenhouse gas emissions" (Global Warming's Six Americas, n.d., *climatecommunication.yale.edu/about/projects/global-warmings-six-americas*).

Ideally, it's essential to identify values that transcend different political and cultural beliefs. Lakoff's (2004) model of political communication establishes a three-level framework for advocacy that is in the reverse order of purpose-driven marketing. The first level is identifying the framework of values communicators are working from. For example, engaging in climate action involves adopting universal values that transcend political affiliation, such as fairness, protection of nature, respect for individual autonomy, future welfare, efficiency, affordability, and waste avoidance (Corbett, 2021).

Media ethicists have identified several "interhuman essences" common in most cultures. These values include accountability, social responsibility, truthfulness, free expression, gender, racial equity, community, respect, reciprocity, spirituality, authenticity, human rights, integrity, nonviolence, dignity, and honoring the sacredness of all life (Elliot, 2009). My preference is to connect ecojustice with universal values. Ecojustice recognizes that environmental predicaments are inseparable from gender, racial, and income inequality (Martusewicz et al., 2015).

The second level involves identifying issue types that connect directly to core values. For example, how might the climate crisis link to the value of protecting nature? How can the problem of climate disinformation undermine one's autonomy to make informed choices? How can human rights or income inequality be tied to climate issues? The third level is the actual policy that addresses these broader concerns (such as taxes, regulation, and cleanup) or actions to be taken (protests, petitions, donations, etc.). There should always be an actionable call.

Climate communicators should be aware of what motivates people to take action. Maslow's hierarchy of needs offers a good discussion point for students to think about what drives their actions and those of their peers. Start by doing a web search for an image of Maslow's hierarchy of needs and share it in class for discussion. Next, ask students if they agree with the assumptions made in the diagram. Here you can problematize this, as it can be considered ethnocentrically based on Western values. For example, Maslow visualized this hierarchy of needs with self-actualization at the top of a pyramid, followed by (in descending order) esteem, love/belonging, safety, and physiology. However, a

Native American critique asserts that humans are born self-actualized, which is reinforced by community actualization for generating cultural perpetuity.

From this Native American perspective, education is a matter of nurturing these qualities because self-actualization is not something to aspire to or is the consequence of some individualized developmental process during a single lifetime. Instead, it results from ongoing generational continuity (Ravilochan, 2021). After this discussion, next, show a product ad or social marketing campaign's PSA and ask students if they can locate on Maslow's diagram how the communicator is trying to activate and motivate the audience. And then see how it fits within the Native American critique.

The credo to know thyself is essential for climate communicators. To achieve this, I have two short writing assignments. The first is a fossil fuel autobiography, which helps students identify the many ways fossil fuels shape their lives and become self-aware of emotional attachments to the status quo. This enables them to communicate better to an audience resistant to change and counters climate disinformation that social and behavioral change is scary. The second short writing assignment is to pen a moral vision statement, which helps students better clarify their values.

Fossil Fuel Autobiography

In this 750-word written assignment, I first ask students to do the online Cambridge Carbon Calculator (any carbon calculator will do) and then to reflect on the results. They are prompted to describe their emotional response to estimating their carbon footprint and noticing what gets their attention, including anything they had not thought about before about their behavior or consumption practices. Next, they write about whether they know the different impacts they calculated and if they could reduce their carbon footprint by 2030 (the calculator recommends reducing it by 10%). What emotions do they have around changing their behavior (if at all)? The next step is to consider the many ways they benefit from the fossil fuel culture that are not covered by the calculator and to consider how much of their lives is a product of the history of fossil fuels.

Some possible writing prompts include: What is the role of consumerism in your life? If you reduce fossil fuel consumption, would it change your identity? What consumer behaviors are you attached to? Students are instructed to pay attention to their emotional responses and to not be afraid to write honestly, even if they think it's not popular or "politically correct."

Moral Vision Statement

To prepare a short 400–700-word moral vision statement, students are prompted to read examples of different environmental mission statements, such as the

Earth Charter (Freeman, 2017) that inspired this assignment). In class, we practice timed automatic writing, an activity in which one writes continuously for a set amount of time (I do it for seven minutes) without pausing. This technique gets past the "editing mind" that inhibits the free flow of thought and association. As part of the instruction, it is important that they strictly follow the rule so that writing never stops. For example, explain that they can write, "I don't know what to write," if they are stuck, but they must keep their hands moving. This warm-up exercise can be done with several prompts (choose one for each time it's done):

- How should humans relate ethically to (and value) the more-than-human world (animals, plants, ecosystems)?
- What kind of post-carbon world do you envision?
- It's 40 years in the future, and we solved the crisis. What did we do?

Next, students are tasked with working on their own to write their personal mission statement, which includes describing the values that inform their relations with the more-than-human world and the environment, a description of what the "good life" should be like, and an action plan for engaging family or community.

Some things to consider when working on the vision statement. What about our identities needs to change (i.e., convenience, disposability, endless growth, and individualism without limits)? How do you address fears that keep people from changing (and holding on to the old system)? When it comes to the identity of consumerism, what are they losing or sacrificing? We are already sacrificing and losing so much; what are they losing anyways? Be open to change: what can we gain by making changes? How will our lives improve?

Once students write these individually, they are tasked with combining them on the class blog website (discussed below) as a joint mission statement. For example, this is what they wrote the last time we did the project:

> Through this blog, we envision a "good life," where people can achieve a sense of contentment through their union with the non-human world and focus on improving themselves instead of achieving satisfaction in accumulating material goods. With our posts, our goal is to get more people to understand the effects and causes of human activities on the environment, hoping to reshape our relationship with it.

Students can advocate converting the system for a real environmental change to happen. Education is an effective resource; when people are more aware of the topic and have a better understanding, they will be more motivated to take action. If we put our minds and hearts toward a new ecological direction, it will be transformed into material action.

Multimodal Media Writing

Once students clarify their values, they engage in several practical multimodal writing activities, each with scaffolding skills for the next project. The first assignment is a class blog, which teaches fundamental writing skills that are utilized in all the other assignments. Next, they create climate memes to engage visual communication and climate discourses. This is followed by an in-class PSA writing exercise where they develop medium-specific writing for an audiovisual script. The final assignment is a written proposal and media kit. Due to the nature of a course focused on writing, I don't have them produce the actual project.

Basic skills are reinforced repeatedly, such as in every assignment, they have to define what they want to communicate in a single sentence. They must also identify the specific audience by age, gender/s, geography, and culture. And, they need to clarify the tone, discursive style, and communication approach (such as serious, casual, humorous, or scary); and apply it appropriately to the medium properties of the project (i.e., blogging, meme, audiovisual script, or digital storytelling).

I also encourage lifelong writing habits. As a journalist who regularly wrote three to four articles a week, I learned to write quickly and efficiently based on several essential writing skills (some of these suggestions are also derived from *The Huffington Post Complete Guide to Blogging*, Editors of the Huffington Post, 2008).

- Pre-write in a journal or notebook and have it accessible to jot down ideas when they pop up (a note-taking or voice memo app also works).
- Use a dashboard and social media management tool to monitor, gather information, and log ideas; clip web links; and draft posts (such as Netvibes, Hootsuite, or Evernote).
- Organize a writing schedule by blocking reasonable chunks of times that fit your rhythm.
- Just write (use the automatic writing technique) and be Zen (don't try too hard, relax). Perfect is the enemy of done.
- Write like you speak (authenticity is important—sound like a human, sound like yourself).
- If you get stuck, imagine you are writing an email to an intelligent friend or relative.
- Focus on specific details that connect with the senses.
- Own your topic and be the go-to expert.

- Know your audience (a blog is only as good as the people who read it).
- Write short.
- Be part of a conversation with like-minded media-makers and bloggers.

Climate Blogging

Blog post composition cultivates fundamental skills applicable to journalism, proposal writing, and media kits. For the strategic climate writing class, students produce a group blog over six weeks that models an actual publication. Using a shared online document (such as a Google Doc or similar platform) as an editorial planning document, they define the blog's purpose with a mission statement (see above) and devise a title and subtitle. They define roles that include managing editors, designers, in-house techies, writers, and a social media manager (see Cruger, 2017 for detailed instructions).

The class blog is broadly themed around climate action, but within that framework, the structure is open to allow students to work through how they want to approach it. Though the parameter can be wide-open, I recommend that they focus on the local student community as its target population. They must think through how to best reach their peers and self-consciously construct a language that best reaches this group.

For example, one of the themes of the editorial meetings was that students wanted to focus on reducing consumerism and waste. As a result, many posts were about where to find used clothes or buy organic food. Though blog post topics are flexible, my experience is that it's better to give them specific limits to work with. Too much choice bogs down the assignment with irrelevant posts (e.g., one student wrote about earthquake preparedness, which didn't make sense on a climate blog).

I suggest they choose from the following kinds of blog posts: expand on an idea and develop an analysis from the required reading; write a media review of a book, comic, film, TV show, or social media channel; introduce a video or podcast and explain why people need to pay attention; write a climate activist profile; analyze a climate image circulating in the media; pen a cultural commentary/critique; discuss and research a climate news event; prepare an explainer (helping readers understand a climate issue); analyze climate data graphics; or discuss, review, and critique media from the class website or shown in class.

Blog posts have three essential elements: title, lede, and body text. In addition, as a form of digital writing, they include tags, hyperlinks, and embedded media. Each blog post should have:

- A compelling headline and a complementary story summary (called a deck or drop head).

- A creative, unique, and professional one-line hook called a lede (or lead).
- At least two body paragraphs (400-word minimum).
- A central discussion with an informed opinion and angle.
- Research that includes links to original sources.
- Embedded Creative Commons media: audio, video, maps, or images with links.
- Call to action (CTA).
- A walk-off line or sign-off.

We use Padlet as a staging area where students can post links to web materials to generate ideas for the blog. In the first week of the blog assignment, students go through a curated list of environmental blogs and choose one site to review as their first post. They must also post a link to their three favorite blog posts on Padlet and then use the commenting tool to respond to each other's selections. Finally, they present their selected blog post for a detailed breakdown in class. This activity gets them thinking about what constitutes strong writing for the medium based on the qualities discussed above (such as the SUCCES model). During class, students work together to plan future content and edit each other's blog post drafts.

When it comes to drafting the blog post, what's more important, the headline or body? Turns out it's the headline: it's the post's elevator pitch. This is one writing skill I spend the most time on because it expresses the post's main idea, prompting the reader to understand why it's worth reading and what value is being added to the topic. Mastering this skill is useful for email outreach (it's the subject line), press releases, crowdfunding campaigns, getting attention, search engine results, and linking on Twitter. Of course, only 20% of people will click through, and the difference between good and bad headlines could be 500% of traffic. In class, I ask students to write 15 different versions of the headline for their first blog post. I then ask them to eliminate ten and then share the top five with a classmate to determine the best headline. We then read them out loud and discuss them during class.

Other skills we focus on are ledes and summaries. The varied methods and techniques for doing this are beyond the scope of this chapter, but any good journalism textbook has exercises that can be used in class. In general, like the SUCCES model, the following tips are important for cultivating storytelling skills and writing capacity for the other projects that follow:

- Start a conversation and offer a new spin so that readers learn something new.
- Write about what people are thinking/talking about and tie posts to a current or recent event.

- A post should aim to make one key point (take-home message)—potentially summarized by the headline.
- The opening line restates or expands on the headline.
- Body proves headline/lede.
- Conclusion restates basic message/take-home message with a walk-off line.
- Use stories and anecdotes.
- Be memorable and sticky by connecting with the reader's direct experience; go for an emotional reaction; use an element of surprise; avoid confusing the subject or writing by sticking to the main point.
- Offer facts, not just opinions.

Most importantly, the writer must be clear about why the reader should care about the topic. Other key takeaways include making the post sound like it is written for the designated audience by connecting with their identity, values, and worldview. Environmental communication fails when it doesn't address emotions. It's important to recognize and promote positive emotions, such as biophilia, love, hope, solidarity, and love of place by engaging with the heart (and not just the head). Emotions can drive and support rational thought through stories that connect experience and feelings with facts, research, and appropriate use of details. I always emphasize that all good writing, regardless of form, is based on extensive research.

These skills are practiced through writing multiple blog posts, which prepares them for their other writing projects.

Memes

The next step is to develop visual writing skills by critiquing and creating memes, a popular form of communication that students enjoy working with. A meme is a form of cultural rhetoric that is shared between people that often performs the function of propaganda (Hobbs, 2020). Meme comes from the ancient Greek word *mimme* (to imitate) and refers to internet artifacts that are short blasts of visual media (video or images) that are shared, copied, and modified (sometimes also called viral media). The purpose of this activity is twofold. First is to learn the art of visual metaphors, the currency of visual culture. The second is understanding how certain climate phrases generate different stories-to-live-by (Stibbe, 2015).

This activity requires two class periods. Introduce the assignment by directing students to the *Know Your Memes* website *knowyourmeme.com* for searching for specific climate memes. They are instructed to post their three favorite climate memes on the course Padlet (the same used for the blog) and to write a very short justification for each choice. In class, every student presents one meme from the Padlet for discussion. After each presentation, students work in pairs to fill out a meme worksheet adapted from Hobb's meme-as-propaganda activity:

1 *Message*: What ideas about climate are being expressed? What visual elements are used?
2 *Context*: What are the hopes, fears, or grievances about climate present in society now? Consider the economic, political, and social environments.
3 *Point of View*: What visual and verbal evidence in the text offers ideas concerning the author's point of view?
4 *Audience*: Who is the target audience? What is it about this message that would be appealing to this group? What reactions might different audiences have?
5 *Creator*: Who is the meme creator? Why was this message created? What do they hope that the audience will think, feel, and do? Many climate memes come from different political persuasions, so it's important to get students to decipher the intent and goals of the specific meme.
6 *Consequences*: What potential beneficial or harmful effects could this message have on individuals and society? Is it an example of beneficial, destructive, or ambivalent forms of climate storytelling? *propaganda.pathwright.com/ library*

Close with a discussion about what makes the meme viral and why someone would want to share it.

The follow-up assignment is to create their own meme. To prepare, I introduce the subject of climate framing and how the terms we use impact how we think about climate change. In using the following list (it can be expanded or reduced), students are asked to discuss which terms best describe their perspective and how each phrase frames a climate worldview:

• Climate change
• Climate crisis
• Climate emergency
• Global warming
• Global weirding
• Global heating
• Climate disruption
• Climate chaos
• Environmental breakdown
• Climate denier
• Climate skeptic
• Climate delayer

They are then instructed to use any of these phrases in a climate meme they produce using a meme generator website. They can choose an existing meme, rewrite the text or image, or create one from scratch. Outside of class, students post their memes on the course Padlet and then present them at the next class meeting. For the in-class follow-up discussion, use these prompts: Which climate concepts apply to the memes you created? What do you notice about your own creative process when creating memes? What visual metaphors are being used and what are their meanings? How do they communicate differently than other kinds of writing?

The memes are then posted on the class blog and its social media channels.

Public Service Announcements

PSAs are short-form social marketing media that utilize audiovisual or audio formats advertising techniques. Here, I focus on the audiovisual PSA. However, it might be considered a relic of legacy broadcast television (traditionally 30 seconds or one minute in length), but newer and longer form PSAs can be seen distributed on social media platforms. For this activity, I work in the more traditional legacy format because it's good to be constrained by the short structure. Students can be assigned to make an actual video, but we focus on the proposal for a writing course. This assignment can also be substituted with a digital storytelling project.

To prepare for the activity, search YouTube for examples (preferably environmentally themed) and show some classic PSAs (such as *This is Your Brain on Drugs* or the *Crying Indian*). The National Ad Council is an excellent source for PSAs. You can also incorporate media literacy activities to teach about persuasion techniques used in advertising and social marketing.

This three-week activity is a good teamwork exercise. After assigning teams (no more than five per team), instruct each team to choose an issue related to the climate concerns of their school community and make a list of possible topics (review, if necessary, the key concepts of purpose-driven marketing discussed above). Next, building on research skills developed from the blogging activity, they compile reliable and current sources on their topic, focusing on the central problem they wish to address. They then concentrate on an action they want to promote that can help solve the problem. Use the following questions to prompt their thinking:

- What would you do if you oversaw a national campaign to fix this problem? List different solutions to the problem.
- Can you imagine what would happen if people started to act differently to solve this problem?

Now they start planning their PSA. To maintain continuity across projects, they should have the same target audience as their blog. They should obtain basic statistics on the school population and decide if they want to focus on students, faculty, staff, parents, or administrators. They should brainstorm the style of communication they want to use and how they will convince people to care about the issue. They choose persuasion techniques and language to invoke emotion by using attention-getting hooks used by advertisers. As with the meme activity, use visual metaphors, such as a frying egg for a brain on drugs used by the famous anti-drug campaign of the 1980s. They should decide what important words, phrases, and/or statistics should appear on the screen and how they will dramatically communicate these facts.

Next, they develop a concept plan and write it out in the form of a proposal (see below), storyboard, and two-column video script (templates and examples are easily available with an online search). The two-column script has audio (what people hear) and video (what people see) in each column. They should use the technical language of film (i.e., LS for a long shot, MS for a medium shot, and so on) to specify each shot (guides are easy to access on the web). PSAs always end with a call to action, which could be anything from directing viewers to a website, voting in an election, signing a petition, attending a demonstration, to changing behavior (like recycling).

Some things to keep in mind. For a PSA, they need to write for both the eye and the ear, although there are examples of PSAs with no sound. PSAs rely heavily on visual metaphors to represent complex ideas in a simple form (such as garbage as a synecdoche of wasteful behavior or a talking jellyfish to represent the sea). Finally, there is a fine line between sensationalism and strategic messaging, so be careful about using strong or violent imagery that could potentially turn off the audience (i.e., showing slaughtered animals to promote animal rights usually doesn't work).

The Proposal

Media teachers such as myself, who were suddenly thrust into remote teaching during the COVID-19 pandemic emergency, could no longer rely on the technology and resources of school facilities, so assigning production projects became difficult. One solution I came up with was to substitute an actual media production assignment with a written version. The results were so strong that I continue to assign the written video project in the form of a proposal as a regular assignment.

Here, I provide the actual instructions for audiovisual project proposals. This template can be used for PSAs, digital storytelling projects, video essays, or other media production projects. Blog techniques they developed from writing titles, ledes, and summaries can be applied to the project title, longline, and synopsis.

Overview: Write a proposal based on your story idea and put the outline into a slideshow that will be presented in class. The presentation should have one slide for each element, accompanied by an image.

Title: By capturing the main concept, the title is a prompt to cue your audience by maintaining a balance between being too general or too specific. Start by listing keywords that summarize your main idea and develop your title based on those. A simple title can also tie into a broader activist branding campaign, like the Rainforest Alliance's *Follow the Frog* project, or the UN climate initiative and eponymously titled website, *Don't Choose Extinction*.

Logline: This one-to-two-sentence summary of the main concept is written in the third person. Avoid jargon and use simple language—a strong logline has a "hook" to stimulate interest (note: the logline is not a subtitle). This should be your take-home message. In combination with the title, someone should be able to read this and get a sense of what the story is about. A great example is from a short film called *Matagi Mālohi: Strong Winds* (*youtu.be/LXpok5rdt6g*): "Matagi Mālohi tells the story of our journey to uplift our people and shape a narrative that paints us not as victims of the climate crisis but as the leaders, the healers, the nurturers, the artists, the gardeners, the growers, the seafarers and the navigators we are." A shorter example comes from the *Follow the Frog* film *youtu.be/3iIkOi3srLo*: "You don't have to go to the ends of the Earth to save the rainforest. Just Follow the Frog!"

Short synopsis: A brief summary of what we will see. Depending on the length of the project, write a one-to-two paragraph description written in the third-person, present tense, with an active voice. Follow the general rule of "show don't tell." What will the viewer expect to see and/or hear? It's not an argument, justification, or opinion about the subject, nor is it the place to describe your thought process. To prompt your writing process, start with, "In this video we see…" (but in your final draft, remove this phrase). Another way to think about the synopsis is that it's like the abstract for your project. It should demonstrate your main assertions and perspective about the topic. As a model, the short film, *A Message From the Future With Alexandria Ocasio-Cortez* (*youtu.be/d9uTH0iprVQ*), uses the following text for its synopsis:

What if we actually pulled off a Green New Deal? What would the future look like? The Intercept presents a film narrated by Alexandria Ocasio-Cortez and illustrated by Molly Crabapple. Set a couple of decades from now, the film is a flat-out rejection of the idea that a dystopian future is a foregone conclusion. Instead, it offers a thought experiment: What if we decided not to drive off the climate cliff? What if we radically change course and save our habitat and ourselves? We realized that the biggest obstacle to the kind of transformative change the Green New Deal envisions is overcoming the skepticism that humanity could ever pull off something at this scale and speed. That's the message we've been hearing from the "serious" center for four months straight: that it's too big, too ambitious, that our Twitter-addled brains are incapable of it, and that we are destined to just watch walruses fall to their deaths on Netflix until it's too late. This film flips the script. It's about how, in the nick of time, a critical mass of humanity in the world's largest economy came to believe that we were worth saving. Because, as Ocasio-Cortez says in the film, our future has not been written yet, and we can be whatever we have the courage to see.

Motive: Be clear about why you are doing this. Why this topic? Why are you the one to do it? Why now?

Target Audience and communication style: Who are you trying to educate and for what purpose? What is the best style to communicate in?

Medium: What form will this take (i.e., photos, video, text, etc.)? Be specific, i.e., TikTok video, Instagram Stories, etc.

Examples of media that you will need: List possible images, sounds, videos, etc., that would be featured in the project.

Venue/distribution: How will people access this? Where will it be distributed, shared, or exhibited?

Treatment: A treatment is a more detailed version of the synopsis. It's a written narrative of what happens in the video, written length usually one-to-two pages, single-spaced. It is not the same as a script, therefore, do not include the actual text of any voice-over (describe only). A treatment focuses on actions and specific descriptions that can be seen and heard, enabling the reader to visualize the project. It's written in the present tense (do not use the future tense, i.e., "it will…") and active voice ("the viewer sees" instead of "the viewer is seeing"). Avoid the use of hyperbole (such as "this unique film will explore"). It is not a rationale for why you are creating it, but a description of what we see (show, not tell). Avoid terms like "therefore," "moreover," and "in addition." A treatment is not written like an essay, so you do not need to use conjunctive adverbs.

The Media Kit

Media outreach skills are not typically taught unless one specifically studies public relations. My personal experience suggests that strategic writing should culminate with a media kit (also called a press kit). Journalists and editors are accustomed to receiving these, either in a physical folder or as online files. Professional media kits use graphic design to highlight important text and to attract attention. The main elements of the media kit are a one-page pitch letter (cover letter) targeted to an editor/media outlet with a clearly stated goal to get some kind of press coverage; a one-page press release; and accompanying materials (biographies, past reviews, press clippings, and/or audio, image, and video files). The media kit should give decision-makers as much information as possible to cover the project. It should be clearly written and uncluttered.

The aim is to make it as easy as possible for the decision-maker to choose the project and assign someone to cover it. The project's communication goal's big picture should be clear: what kind of media coverage should the project garner to reach its target audience? Where should people read or hear about it? And how will that achieve the particular aims of the project? For example, what are possible venues to publicize if students are trying to reach the university population? It could be a student newspaper, university social media channel or website, alumni newsletter, campus radio station, etc.

Once the proper media outlet has been designated, students should research the contact information and correct spelling for the outlet's editor, writer, or manager and determine what outcome they want (coverage, interview, written article, etc.). Importantly, the cover letter is not pitching an announcement, listing, or requesting sponsorship but advocating for an actual article or feature story. Just as they learned from blogging, students should identify strategic news hooks to garner the media contact's interest.

Pitch Letter: The pitch letter should be addressed to a specific media outlet and staff with a clear goal written in a formal business letter format. The letter's tone should be in the style of the publication, finding the right balance of professionalism and appropriate tone. The letter should say exactly what you want and explain why they should do it (a call to action). Use writing skills to convince the decision-maker to cover the project by demonstrating relevance to the immediate community and the significant impact it will have by developing a vivid picture of the project. For example, "The student paper has recently committed to reducing its carbon footprint. A feature article about our blog will let readers know how students can also reduce their ecological footprint. Highlights include…"

Press Release: A press release is a single sheet document that has a headline and sub-headline (18 words or less), one sentence pitch ("grabber"), a short one-to-two paragraph overview that has the most pertinent information; clear dates, and times (if there is an event); an about section; and contact information. It should contain a catchy title, news hook, and relevant statistics (if necessary). The main body should start with who, what, where, and when. Press releases often contain quotes, which journalists can use when they write their articles. Luckily, electronic documents allow for links to be embedded, so the document can be used to direct the reader to more information.

Accompanying materials: A background/bio page supplements the press release with additional information (biographies, quotes, organization profiles, or statistical and factual information) that supports the main communication goal.

Conclusion: Confronting Barriers to Climate Communication

The climate crisis (and solutions) destabilizes our worldview and what we consider "normal." This leads to irrational discounting and pushing solutions to the future: it's someone else's problem; it's too difficult; I give up; it's not my responsibility. The fossil fuel industry is aware of this tendency, so they push climate disinformation that promotes climate inaction by scaring people about disruptions to the status quo. Furthermore, there is the general problem of symbolic annihilation: most media avoid dealing with environmental and climate issues, which is part of historical ecological amnesia and alienation. This requires climate communicators to name and define problems.

The major challenge is that the climate crisis is not a scientific problem but is cultural, political, and psychological. The information deficit model in information and science literacy usually doesn't work because inaction is not based on a lack of information—it's the inability to deal with our information. Climate change communication is then about the individual interacting with their surroundings. Individuals are social (families, society, friendships), cultural (pop culture and entertainment, cultural values, and practices), and embedded in the macro level (political and economic spheres, institutions, and physical environment). Individuals do not exist in a vacuum but through holistic interactions within these various spheres. The task is to connect climate to these overlapping realms of experience (Corbett, 2021).

The climate crisis is traumatizing, which leads to avoidance and silence. What can we learn from COVID-19 to deal with trauma and destabilization? We can deploy writing and storytelling to challenge the dominant narrative, shift cultural trauma into social change, and develop a coherent narrative of climate action. Katharine Hayhoe's (2021) model of rational hope can counter climate disinformation and fear: accept that we are in a crisis and that we must confront it; love Earth and envision positive, healthier, changed ways of living; recognize that our choices matter now and for the future; that cultures are capable of profound and rapid change when there is a crisis; and communication drives these changes.

Emma Marris' (2020) *t.ly/15_E* advice for dealing with climate hopelessness is prescient to focus students on how to apply their climate storytelling skills for life beyond the class: ditch shame; focus on systems (not yourselves); join an effective group; define your role; and understand what you are fighting for (not just what you are fighting against). Don't fight existing reality but create new stories-to-live-by that make the old stories obsolete, i.e., "system change, not climate change." Now, get those writing instruments out and start scribing the future!

References

Corbett, J. B. (2021). *Communicating the climate crisis: New directions for facing what lies ahead.* Lexington Books.

Cruger, K. (2017). The student-run environmental communication blog. In T. Milstein, M. Pileggi, & E. Morgan (Eds.), *Environmental communication and pedagogy* (pp. 244–247). Routledge.

Damico, J. S., & Baildon, M. C. (2022). *How to confront climate denial: Literacy, social studies, and climate change.* Teachers College Press.

Editors of the Huffington Post. (2008). *The Huffington Post complete guide to blogging.* Simon & Schuster.

Elliot, D. (2009). Essential shared values and 21st century journalism. In L. Wilkins & C. G. Christians (Eds.), *The handbook of mass media ethics* (pp. 71–83). Routledge.

Freeman, C. P. (2017). "Moral vision statement": Writing assignment instructions for students. In T. Milstein, M. Pileggi, & E. Morgan (Eds.), *Environmental communication and pedagogy* (pp. 209–211). Routledge.

Global Warming's Six Americas. (n.d.). *Yale Program on Climate Change Communication.* Retrieved July 11, 2022, from https://climatecommunication.yale.edu/about/projects/global-warmings-six-americas

Hayhoe, K. (2021). *Saving us: A climate scientist's case for hope and healing in a divided world.* One Signal Publishers.

Heath, C., & Heath, D. (2008). *Made to stick: Why some ideas survive and others die.* Random House.

Hobbs, R. (2020). *Mind over media: Propaganda education for a digital age.* W. W. Norton & Company.

Kellner, D., & Share, J. (2019). *The critical media literacy guide: Engaging media and transforming education.* Brill Sense.

Lakoff, G. (2004). *Don't think of an elephant!: Know your values and frame the debate.* Chelsea Green.

López, A., Piotto, L., & Macias-Gutiérrez, E. (2020). Integrating information literacy in a communication writing course. In M. Stöpel, L. Piotto, X. Goodman, & S. Godbey (Eds.), *Faculty-librarian collaborations: Integrating the information literacy framework into disciplinary courses* (pp. 59–75). Association of College and Research Libraries.

Marris, E. (2020, January 10). Opinion: How to stop freaking out and tackle climate change. *The New York Times.* https://www.nytimes.com/2020/01/10/opinion/sunday/how-to-help-climate-change.html

Martusewicz, R. A., Edmundson, J., & Lupinacci, J. (2015). *EcoJustice education: Toward diverse, democratic, and sustainable communities* (2nd ed.). Routledge.

Ravilochan, T. (2021, August 18). Could the Blackfoot wisdom that inspired Maslow guide us now? *Medium.* https://gatherfor.medium.com/maslow-got-it-wrong-ae45d6217a8c

Stibbe, A. (2015). *Ecolinguistics: Language, ecology and the stories we live by.* Routledge.

10

THE LONG HAUL

Three Decades of Teaching Student Documentary Action Research for Environmental and Climate Justice

Steve Goodman

Students often view rising oceans, melting polar ice caps, and other disastrous effects of global climate change as disconnected from their lives, distant threats occurring in remote parts of the world. Some may feel that the problems are too overwhelming and that nothing can be done. Educators, particularly those teaching in diverse urban school districts, can play a critical role in helping students connect climate change to their daily lives by using media production to explore urgent questions of environmental science, public health, and racial justice. For example, students can investigate how climate change's catastrophic effects, such as massive heat waves, floods, and hurricanes, amplify pre-existing social, economic, and racial inequities for Black, brown, Indigenous, and poor white communities. They can then research problems, render stories, and propose possible actions at the intersection of environmental and climate justice.

This chapter discusses the collaborative action research and storytelling embodied in four student documentary projects about environmental justice problems faced by New York City communities spanning the late 1980s into the 2010s, and how these critical practices can inform student environmental media projects today. The youth co-created their documentaries in intensive, credit-bearing, after-school workshops at the nonprofit Educational Video Center (EVC) *evc.org*, of which the author was the founding Executive Director. They were drawn from New York City high schools for recent immigrants and small schools serving overage and under-credited students struggling to graduate.

The analysis of these four cases includes descriptions and dialogue from the footage and links to two short clips from one of the documentaries. In addition, the voices of two former EVC Documentary Workshop Program Directors,

DOI: 10.4324/9781003335276-14

a former EVC student and her mother, are woven throughout the discussion. Interviewed while watching back the documentaries they facilitated and participated in years before, their reflections (lightly edited for clarity) provide unique insiders' perspectives on the teaching, learning, and production of these projects.

The first of these projects was produced when the climate crisis had yet to become fully part of the national conversation and before cell phones and social media had been invented. Now that students are all carrying a camera in their pocket and can upload their media messages to the world, the potential for student environmental media projects is nearly limitless. However, the focus is not on the production and dissemination technologies but on the teaching, learning, community partnerships, and actions that these student productions make possible.

Across their four projects, the students focused the viewers' attention on environmental and public health problems that were often invisible to the general public. Advocating for more sustainable and just environmental change, the youth producers documented the deadly chemicals dumped into our rivers and ocean by corporate polluters, *New York City and the Hudson River: Downstream and Up the Creek evc.org/product-page/new-york-city-the-hudson-river* (Educational Video Center, 1989); the tons of garbage carted away each day by trucks and barges, *Trash Thy Neighbor vimeo.com/175876642* (Educational Video Center, 1990). Youth producers also created videos about the diesel fumes of garbage trucks choking poor neighborhoods, *Shame on You! That Can Be Reused! evc.org/product-page/shame-on-you-that-can-be-reused* (Educational Video Center, 2007); and the lead dust and toxic mold growing behind the walls of public housing, *Breathing Easy: Environmental Hazards in Public Housing vimeo.com/66192997* (Educational Video Center, 2013) (for video clip transcripts of these two videos, see *t.ly/L-J4iii*).

Considering all four films together, a larger picture emerges from the cumulative products of capitalism's toxic and unsustainable growth, unbridled consumer excess, and the systemic racism that causes historically marginalized communities of color to suffer the worst consequences. Yet, with a sense of urgency and also humor, these students, family, and community voices remind us that these conditions of environmental injustice are not unalterable facts of life but are artificial problems that can be resisted and changed.

Student-Centered and Inquiry-Based Teaching within a YPAR Framework

Core to EVC's documentary pedagogy was its student-centered, inquiry-based approach and Youth Participatory Action Research (YPAR) framework. This meant grounding projects in students' questions, interests, and experiences. Their home, school, and neighborhood were often sites for investigation, media making, and action. As students tapped their own family and community's "funds of knowledge" (Moll et al., 1992), the traditional teacher's position as the sole source of knowledge was de-centered.

Their student-centered approach strengthened students' accountability and ownership of their learning. For example, before they began their projects, students collaboratively created group agreements to establish norms and ground rules for safe and inclusive conversations and work together. They also drew production calendars and planned backward from the final screening, marking deadlines for the research, shooting, and editing phases that led up to it. The teachers integrated social-emotional learning into the projects by starting workshops with regular team-building rituals such as "highs and lows" (also known as "roses and thorns"). They provided everyone a dedicated time to share personal successes and problems they may have experienced in or out of school. They also fostered students' self-reflection through daily journaling time by asking students to consider what they've learned through their day's work and what creative ideas they may have for upcoming work, such as shot lists, interviews, or edit plans.

While the instructor's facilitative role remained crucial, the routine use of peer teaching strategies reinforced and distributed knowledge as students coached their peers on skills they had just learned to perform, such as operating the camera, the boom mic, or editing software. These authentic learning experiences allowed students to practice new identities as critical researchers, camera operators, storytellers, and environmental educators and activists when they used their videos to inspire viewers to question, debate, and take action in response.

Driven by students' questions, each video production was like peeling an onion; each question and layer of research revealed another beneath it. As former EVC Documentary Workshop Program Director Dave Murdock (April 18, 2022) described the investigative process with the *Trash* documentary:

> If you can start small with like a paper bag blowing down the street and then you ask, what happens to this thing?… you're incorporating daily life. Keeping it small at the beginning. And letting it build up. Once you're on that detective story of, 'What happens to this stuff?' it's really interesting. You start seeing it everywhere. You start seeing the garbage man, and the garbage trucks. You wonder, where do these trucks go?

When her students had questions about the impact of truck pollution in the community, former EVC Documentary Workshop Program Director Christine Mendoza prompted her students to take the next step in their *Shame on You!* project and interview people on the South Bronx streets about the high asthma rates there.

Students developed a sense of agency to question people on the street and the city's water and garbage sanitation officials. Developing their questions and conducting interviews sharpened students' research skills and social-emotional learning as they had to leave their comfort zone and work collaboratively to overcome obstacles and solve real-world problems that arose in the media-making process.

Not only did the EVC instructors weave student-centered and inquiry-based teaching strategies throughout the production process, but they challenged their students to apply their findings to the environmental problems that they were investigating. This approach was informed by the Participatory Action Research pedagogical framework that has long been practiced in Africa and Latin America. This framework is based on the notion that problems of injustice can best be solved by those most directly impacted by them in collaboration with outside researchers. It engages young people as active researchers and draws upon their experiences and insider perspectives to initiate social change and solve the problems that are obstacles to their well-being (Cammarota & Fine, 2008; Goodman, 2018; Petrone et al., 2022).

In these intergenerational "contact zones" (Fine & Torre, 2019) of dialogue, analysis, and reflection, the students drew upon a diverse range of collaborators for their documentary research, including peers, family members, environmental and public health advocates, doctors, scientists, and residents in the community. As co-creators of media art and new knowledge, the students used interviews, verite footage, visual documentation, art, music, puppets, diaries, surveys, games, humor, dramatic re-enactments, and more to re-present their environmental findings and narratives for youth, school, and community audiences. In this way, students' participatory research can redefine what counts as knowledge in a school's curriculum and instruction (Caraballo, 2022).

Student Engagement: Meeting Students Where They Are

Some EVC students initially struggled to connect to the environmental subject of their documentaries. They were also concerned about how their documentary would engage their urban peers in an environmental subject.

Growing up in New York City in the 1980s, EVC students encountered multiple epidemics in their communities, including AIDS, crack, homelessness, youth crime, and police violence. They said they hadn't really thought the environment was an important social problem, at least not one that urban students of color needed to be concerned about.

However, the environmental justice movement had begun to take root in the early 1980s during protests in North Carolina against the state's dumping of the deadly poly chlorinated biphenyl (PCB) chemical in a landfill in a poor Black farming community. It was sparked by the civil rights leader Rev.

Dr. Benjamin Chavis when he declared, "This is environmental racism." He defined it as, "racial discrimination in the deliberated targeting of communities of color for toxic waste disposal and the siting of polluting industries" (Chavis, 1993, p. 3) and "the systematic exclusion of minorities from environmental policy making, enforcement, and remediation" (Almassi, 2021, p. 22). Chavis coined the term and helped bring a racial and social justice frame to this growing movement that was also protesting petrochemical and lead poisoning in poor and segregated communities across the country and the world (Fears & Dennis, 2021; World Economic Forum, 2020).

But environmental racism and environmental justice had not yet become popularized among the youth at EVC. Even though they didn't necessarily perceive the relevance of investigating environmental pollution in their urban communities, as Mendoza (March 10, 2022) put it, "The key was meeting the students where they are. If they're not interested in it, figuring out why they're not interested in it. Finding out what a potential entry point could be with them." Murdock also looked for possible connections and entry points and found one in his students' love of horror movies. Seeking to connect with their urban peers who would be watching their film, *Trash Thy Neighbor evc.org/product-page/trash-thy-neighbor*, they used creativity, humor, and their knowledge of horror movies.

The students opened their video with a low tracking shot, following behind a student's boots walking through garbage-strewn streets of Manhattan, forcing the viewer to take a closeup look at the garbage we throw out, but it never really disappears. An organ ominously plays classic horror movie chords as the unknown walking student narrator in *Trash* (1990) warns us, "There's a danger out there, lurking, waiting, building every single day.... Soon, IT may take over the whole planet.... No one can escape IT, whether you're Black, White, Hispanic, Asian, young, or old. Especially you!" The camera tilts up to a medium shot, so we finally see who is speaking. "IT is –" and in a nod to then-President George W. H. Bush, the student narrator says, jabbing his finger at the camera with each letter, "Read My Lips! G-A-R-B-A-G-E!" At this reveal, a student then gives a piercing horror movie scream off camera as the narrator walks off-screen, leaving a bag of garbage behind in the background.

Nearly 20 years later when Mendoza's video workshop began working on another environmental documentary about the problem of garbage, *Shame on You! That Can Be Reused*, her students similarly didn't initially understand how this problem connected to their lives as urban youth of color. Seeing their resistance to the topic, she recalled thinking, it would be great for my students to meet other young people "who look like them, who act like them and who might influence them to show them that this is an issue that is important for them to explore. So that peer-to-peer interaction and connection were critical in getting them on board to be excited about this project."

They visited and filmed a discussion with an after-school environmental justice youth group in the Hunts Point neighborhood of the South Bronx. The youth were organizing against the truck traffic and the toxic diesel fumes they spread in their neighborhood. One of the youth participants explains, "We do social justice and environmental justice... It's not fair. Everybody should try and take care of their environment equally instead of always dumping everything on the South Bronx."

The environmental justice youth group was a big inspiration for the EVC crew. They helped shape the direction the video crew would then explore, and environmental racism became a central focus of their film. In their opening narration, the EVC students state, "Environmental racism affects a large number of people who live in low-income communities. While making this documentary, we went to the South Bronx to investigate the injustices in the area."

Even as the EVC students came to learn about how important the concept of environmental racism was, they knew they needed to find an engaging way to teach their peer audiences about it. Mendoza (March 10, 2022) explained that "the students said, 'If we're going to make this film, it has to be fun and it has to be exciting. And we have to have fun making it.'" Like the *Trash* crew, these students got creative. They acted out playful scenes about recycling in the school lunchroom, staged a student recycling game, and made two puppets (whom they named Billy and Betty) out of paper bags to move the narrative of their documentary forward.

Mendoza (March 10, 2022) felt that it was important to support her students' puppet making and other creative ideas for their project and said, "'How can we make this work?' was always the question." While designing, scripting, voice acting, handling, and shooting the puppet scenes, they start with puppet Billy coughing and breathing in an inhaler. Betty asks, "What's wrong?" Billy says, "All the garbage transfer truck fumes are making my asthma act up." Adding statistics to the scene Betty says, "One in four people in the South Bronx has asthma. Can this be the reason?" A title card then appears: "It is estimated that around 3,000 trucks drive through Hunts Point every day."

The disproportionately high rate of asthma in low-income communities of color was also a focus of the 2013 EVC documentary, *Breathing Easy: Environmental Hazards in Public Housing*. In this case, the students investigated the lead poisoning and toxic mold epidemic in New York City's 400,000-resident public housing system. Although they knew that reporting on these hazards would be of general interest, to engage student audiences effectively, they wanted to personalize the problem and document how it affected another student.

One crew member, Raelene, had described to the class her family's unhealthy living conditions and her struggles with asthma. However, she was initially not comfortable being filmed for their project. She finally agreed after her peers convinced her that her participation could help others who may not be aware that they are living with lead, mold, or other similar hazardous conditions in their homes. So she took a video camera home and recorded her thoughts in a video diary.

Starting the Inquiry: Defamiliarizing the Familiar

A strategy that the EVC instructors used to help their students connect with the subject of their films was to teach them to take notice of that subject in their daily life. To do so, students observe the ordinary things they may have passed by or stepped over before with the naked eye and then use their camera to portray these things. In other words, they were learning to defamiliarize the familiar everyday experiences. As a result, across each project, the students became more aware of what they might take for granted and began to ask questions.

In the students' opening narration for *New York City and the Hudson River*, they call upon their viewers to take notice of the familiar, "You know when you cross the George Washington Bridge on the way to New Jersey? Ever notice that body of water that you pass over?" A wide shot captures the George Washington Bridge, spanning the river beneath it. Murdock (April 18, 2022) says:

> You have this huge river in your city. Did you ever think about it when you're driving over? It's getting them engaged in a personal way on the thing... questioning about how clean it is. Or why isn't anybody swimming in it, or can people fish in it. It's kind of opening your eyes to what's right in front of you.

The students working on the *Shame on You!* documentary also learned to make visible the seemingly invisible problem of the deadly polluted air that was all around them. In this case, it was the diesel exhaust fumes and fine particulate matter from sanitation trucks and the nearby highway traffic. Even though it was causing many residents in the South Bronx to suffer high rates of asthma, the heavy traffic was often seen as just an immutable fact of life.

As with the *Trash* and *Hudson* projects, Mendoza also helped her students to make personal connections with their documentary, *Shame on You!* (2007). She helped them to document and make the problem visible by surveying residents on the streets of Hunts Point about whether they knew anyone with asthma.

Mendoza helped her students prepare for their first "shoot" in the Bronx neighborhood by role-playing mock interviews in the classroom and debriefing what they learned from the experience. While some students played the interviewee, others practiced how to introduce themselves and their project to

VIDEO CLIP 10.1 Clip from *Shame on You*. Online available at: https://vimeo.com/ 369988483

a stranger, hold the microphone, actively listen and ask follow-up questions, properly frame the camera shot, monitor audio levels, and communicate with each other as a team.

When they arrived in the Bronx, one of the first people they interviewed said, "Yes, my daughter has asthma." Another man said, "Yes, my mother, my son." A woman answered, "Including myself, I have my granddaughter. And I think the majority of the people who live here." Her estimate was likely not too far off since according to the students' research, the South Bronx asthma rate was three times the national average (see Video Clip 10.1).

Mendoza (March 10, 2022) described her students' reactions after they gathered this survey data:

"Man! All of these people in the South Bronx have asthma.... I didn't ever think it was connected to the environment that I was raised in. OK, maybe this is a bigger problem than we had originally thought....We found this! Now we have to find someone who's going to tell us about this!" So, it was like following the students' curiosity.

When they started to defamiliarize the familiar everyday sight of trucks passing through the neighborhood, they could begin to analyze the connections between the traffic and the rise in respiratory illness. In fact, they found reports that documented two to three truck trips per minute hauling garbage in and out of the residential streets and data that showed the South Bronx has the highest age-adjusted asthma death rates by far among all counties in the state.

In *Breathing Easy*, the filmmakers also called attention to and made visible hidden environmental hazards of lead dust and mold not only for their viewers but also for EVC student Raelene and her mother, Michele Holmes. Michele (March 25, 2022) explained the hidden nature of the mold after the New York

City Housing Authority (NYCHA) moved her family into what they claimed was a mold-free apartment in the building:

> I didn't know what to look for... I didn't know it was already there and they just painted over it. And after being there one month, it was my first ER visit in many years. I didn't know it was because of the mold. I didn't know what those spots were. And they were appearing all over the bathroom, which was connected to my bedroom.... By then I already had Raelene, she had her first asthma attack a couple of days after she turned five. Later on one of my grandsons, he developed it. A couple of days after he turned four. And the common denominator was we all slept in my bedroom. The three of us. And to this day we still have it.

Following the Inquiry: Environmental Racism and Public Science

As Murdock (April 18, 2022) had observed, once his *Trash* documentary students started noticing the garbage trucks all around the city, they asked, "Where do these trucks go?" Following the trucks, they ended up atop the Fresh Kills landfill in the borough of Staten Island. The sanitation worker they interviewed for *Trash* (1990) explained, "This is the biggest landfill in the world and the only operation where the garbage comes in by barges." Their camera pans across cranes and bulldozers, dwarfed by mountains of garbage as far as the eye can see, as swarms of seagulls circle and dive around them.

At the time, dump trucks and barges were bringing in about 29,000 tons of garbage daily. Before it closed, 150 million tons of waste had been dumped there (New York City Department of Parks and Recreation, n.d.). This waste is the byproduct of our throwaway society, where we recycle just 5% of our plastic waste and throw out 15–25% of the food we buy, releasing methane into the atmosphere (Gammon, 2022; Lippard, 2016).

When the Fresh Kills landfill was finally closed a decade after the crew filmed there, the city's solution to disposing of the unsustainable flood of residential garbage was to ship most of it into low-income communities of color for processing by multiple privately operated transfer stations. So, in 2013, when the *Shame on You!* documentary crew also followed a trail of trucks and garbage, they broadened their inquiry to make visible the hidden forces of environmental racism and the social inequities that impacted Black and brown residents in one of the poorest communities in the Bronx.

To understand the concept of environmental racism, students not only needed to learn about pollution and public health but also about the social, political, and economic dimensions of how power works in their city. Whenever possible, the EVC instructors connected these issues to students' lived experiences and to more complex political and scientific concepts explained by scientists, doctors, economists, and public health advocates. In their video, the students made these more abstract concepts easier for their viewing audiences by editing in more plainspoken comments from community members and concrete images that illustrated their meaning.

EVC instructors facilitated other students' critiques and used professionally produced documentaries to teach these advanced editing techniques. Sometimes by showing clips with the audio off, they analyzed how editors used images, graphics, narration, and sound bites to tell stories and explain complex ideas. Students also sharpened their editing skills when they presented their works-in-progress in rough cut screenings that the instructor facilitated with tuning protocols. Giving and receiving multiple perspectives of critical feedback from peers, teachers, and project partners helped students improve their videos and deepen their understanding of how audiences make sense of their work.

In *Shame on You!* (2013), a community development leader explained the concept of environmental racism as follows:

> The communities with very little economic power, almost always low-income, get stuck with polluting facilities, whether it is a bus terminal, a sewage treatment plant, or a garbage incinerator, and things like this. They are put there because the communities are poor, they have very little political clout, and land is cheaper there than in any other part of the city.

To express the political and economic concept in everyday language, the students edited a sound bite of a Bronx resident who put it plainly, "Manhattan is money. They don't have to deal with the circumstances that we have to deal with, 'cause we don't have that many high-class uppity, uppity people like that running around." Community activists further reinforced this idea from a green worker cooperative and from a public housing association who told the students, "They dump on people of color. We've all been dumped on for many, many years. It's an injustice. It's environmental racism." And finally, the students learned to illustrate important ideas from these interviews with b-roll images of trucks driving through neighborhood streets and of a cement mixer and power plant. As Mendoza (March 10, 2022) explained:

> There's something very powerful and authentic about hearing a professional ... say something, and then having the students go out and see it for themselves .... And every single time we did a street interview, we made sure that the b-roll shots and background were reflecting what we were talking about. ... visual storytelling was important in those street interviews.

In *Hudson River* (1989), the students began their inquiry by talking with the men who were fishing on the river banks in Harlem, many to bring food home for their families. When asked whether the fish they caught were safe to eat, one of the men assured them, "You could tell that the fish are clean by looking at the gills…. When you catch a sick fish, it's very obvious."

Interviews with environmental scientists and advocates helped deepen the students' scientific knowledge base and develop their curiosity and scientific inquiry habits. For example, when they interviewed the Hudson Riverkeeper, they learned that in fact, you can't tell whether a fish is contaminated with PCBs just by looking at it, and the chemical is dangerous in very small amounts.

"It's another detective investigation," observes Murdock (April 18, 2022). "You have to check out if you can really tell if the fish are sick. No, you can't. PCBs, you can't taste them, you can't smell them. Well, what's a PCB? And then you're onto this next question to find answers to."

The *Hudson* video crew interviewed an environmental scientist at Columbia University to learn about PCBs. They learned that PCBs were an industrial chemical used to manufacture transformers, capacitors, and other electrical equipment. Moreover, one of the major electrical manufacturing plants in the country was General Electric, located on the Hudson River 150 miles north of New York City. Between 1942 and 1977, General Electric poisoned the river, dumping an estimated 1.3 million pounds of toxic PCBs into the water (Riverkeeper, 2009). From their research, the students included this warning in their film's (1989) opening narration: the Hudson River was "so polluted that New York State warns that if you eat more than one fish per month, you could get cancer."

Although their cameras and the documentary they were producing gave EVC students access to people, locations, and institutions other teenagers would not normally have had, they still felt intimidated going into an elite university to interview a scientist. Recalling the interview, Murdock (April 18, 2022) advised, "They might think, 'I don't want to talk to him, I don't know anything about the Hudson River.' Well, that's why we're going to talk to *him*. You don't have to know everything about the river when you talk to them— you have to do a little bit of research to ask the questions. But generally, you're there to find out from the interview, and then you take that interview and go somewhere else with it."

Their inquiry led them to research what is being done to prevent these toxic chemicals from polluting the river. They visited the North River Wastewater Resource Recovery Facility in Harlem, which treats the wastewater before it flows into the Hudson. They were angry to learn that when storms flooded the city's sewers and drain pipes, the Facility released untreated household and industrial sewage and toxic waste into the river.

Students in the *Breathing Easy* project also had to learn about, and then explain to their audiences the hidden dangers of chemical hazards for vulnerable populations. While it was important to learn the science, Raelene (March 25, 2022) recalled how challenging it was to understand the medical language a doctor used to describe the negative effects of lead in the blood, "The doctor, just every word that he spits out, I was like, 'What? Are these clinical terms? Like, dumb it down a little bit!' So his section was a little hard to edit." But the students made the potentially lethal impact of lead poisoning—especially for children—easier to understand with title cards of facts from the CDC. Moving from research to action, they next edited a scene of an environmental health advocate showing Raelene's mother how to test for lead dust in her apartment.

Taking Action: YPAR and Community Partnerships

Community partnerships with environmental activists and funders were essential for all four student documentary projects. Following a YPAR approach, EVC instructors introduced the youth to the rich and diverse assets within their community, including environmental justice educators and advocates, researchers, and public health experts, to teach them about the problems they were exploring.

However, *Breathing Easy* was the only project where a community-based environmental justice organization also advocated for one of the students and her family, who were directly experiencing the environmental racism that the students were researching. This was only possible because Raelene and her mother Michele decided to allow the EVC students to document their story. Although it was not an easy decision to make, Michele (March 25, 2022) reflected on the conditions that prompted her to open her home to her daughter's class video project:

> Well at that point, I was at my wit's end.… I was angry, but I was tired. I didn't know what to do. So, at this point, it was like, maybe this will help. At that point, I blamed myself for the mold. I blamed myself for my daughter and one of my grandsons developing asthma because of the conditions we were living under. The asthma attacks would occur every month after I cleaned the mold in the bathroom and sprayed down the entire kitchen for the roaches. And I was just at my wit's end.

What made this collaboration with WEACT for Environmental Justice a transformative experience for the youth production team, and particularly for Raelene and Michele was the close relationship they developed with the organization's environmental justice staff, particularly with the community health advocate Ana Parks. Significantly, this was an intergenerational family partnership where both Raelene and Michele felt heard and supported, giving them social capital for what grew to become a long-term relationship with WEACT. As Raelene (March 25, 2022) reflects:

> I will never forget that walk with Ana Parks that we have at the end of the [film's] intro. Because I had never really talked to anyone like Ana before. I never really found myself pouring my story out to anyone.... I connected with her more because she stuck with us. She fought with us.... So there was a real relationship there. So the fight continued with her.

The class collaboration with Ana Parks and WEACT also deepened Raelene's consciousness of how these environmental hazards impacted her health, which later led to action. She recorded her reflections on the "silent killers" of mold in her home in her video diary. Watching her video diary back ten years later, Raelene reflected on her process:

> I had no sense of direction for this diary. I just figured, you know what? I'll treat it like an actual diary and talk freely and see where that takes me. I thought... they're lurking in your house and the next thing you know, you're very sick. And you don't know why (March 25, 2022).

Raelene's conversation with Parks and a section of her video diary was edited into the documentary's opening sequence, "My Ceiling Looks Ridiculous," and can be accessed here in Video Clip 10.2.

When Parks inspected their apartment, she was shocked by the extent of the mold infestation covering the bathroom ceiling. She noted her findings, took samples for lead dust testing, and explained the next steps she would take in

VIDEO CLIP 10.2 Clip from Breathing Easy documentary: "My Ceiling Looks Ridiculous." Online available at: https://www.tcpress.com/goodman-chapter-one

VIDEO CLIP 10.3   Clip from Breathing Easy documentary: "I Don't Know Where to Turn." Online available at: https://www.tcpress.com/goodman-chapter-one

response. Parks' home inspection and dialogue with Michele modeled for the student producers and their viewers how they too could use their cameras to document the problem, partner with community environmental activists, and advocate for their families if they found lead, mold, and other environmental health hazards in their homes (see Video Clip 10.3).

Of course, Raelene and Michele were not alone in their struggles with the city government for safe and healthy living conditions in public housing. In fact, health hazards were rampant and systemic in the aging buildings run by NYCHA, putting thousands of low-income residents at great risk. NYCHA violated the Americans with Disabilities Act for its failure to remedy the mold for residents with asthma, and other federal and local laws when it failed for years to inspect for lead-based paint in its apartments. As of 2022, forty billion dollars in repairs are still needed in NYCHA, the biggest landlord in New York City, and the largest public housing authority in North America. In addition, Hispanic and Black female-headed families made up 93% of all families in NYCHA public housing. Given the scope and racial inequity of the problem, the EVC students understood this as a case of systemic environmental racism (Goodman, 2018, p. 21).

Moving from research and documentation to action took a range of forms and necessarily looked different for each project. Student action can differ in YPAR projects depending on their goals and reasons for conducting their research in the first place. Some result in a conference presentation, a co-authored policy paper, or even marches and sit-ins (Petrone et al., 2022).

Each EVC youth producer presented their environmental documentaries at premiere screenings with school, family, and community audiences for public dialogue and action. They also won multiple awards and were featured at national and international youth media festivals and, by the early aughts, were also disseminated online. In addition, students were invited to present them at various libraries, colleges, nonprofit organizations, and funders. More specifically, *Hudson River* was broadcast as part of environmental programming on WNET public television. *Trash* was screened by environmental partner Christodora. *Shame on You* was used in schools and communities for environmental education by the partner Council on the Environment of NYC (CENYC). EVC students and the Holmes family presented *Breathing Easy* at WEACT's national housing conference, and the film was exhibited at the David J. Sencer CDC Museum in Atlanta in association with the Smithsonian Institution.

Students also modeled and documented actions within the production of their documentaries. For example, the *Trash* student producers not only made us aware of community buy-back recycling centers but also used the final segment to show how they can recycle and reuse instead of throwing more garbage in the landfill. A student spills a drink on a table and uses a sponge instead of a paper towel to clean it up. Another student, with his face completely lathered in shaving cream, leans right into the camera to tell us to use razors where just the head comes off instead of using a fully disposable razor. He then flips the head off and says, "Capisce?" After watching these scenes, Murdock (April 18, 2022) said, "The thing that's really important, on both the *Hudson River* and the *Trash* videos, is the humor. It allowed them to be funny and to find something that's funny in it."

Teaching young people how to recycle was a goal promoted by CENYC, one of the partners for the *Shame on You!* documentary. The crew interviewed a teen who was actively improving recycling efforts in his public housing complex. In addition, the EVC students staged and recorded a recycling game in the middle of their high school hallway to teach their peers how to recycle bottles, cans, and plastic containers properly. Mendoza (March 10, 2022) remembered:

> Students thought that no one would want to play the recycling game. So they thought—"What can we do to entice people to play the recycling game? Let's give them food! Teenagers always want food!" They baked their own cookies and brought them in the next day. But they were so engaged, none of the people who played the game even took a cookie!

In the *Hudson River (1989)* project, the students' main goal was to develop their viewers' civic engagement and environmental awareness of corporate and government malfeasance and accountability in the face of massive degradation of our rivers and oceans. They learned that New York City dumped almost five million tons of untreated industrial toxic sludge in the ocean each year. The students end their film with a closeup of the Riverkeeper standing on the banks of the Hudson, strongly urging citizens not to allow or make an exception for those who break environmental laws and degrade the natural resources that belong to all of us. "The fish belongs to the people of New York. The water belongs to the people of New York.... If we can send that message to these people, that those are our standards, it's going to have an effect on the rest of our lives and the rest of the things that we live with." The frame freezes and fades to black.

The message here is, as Murdock (April 18, 2022) puts it:

It's not a natural force. It's all human decisions. We've made the decision somehow not to put a lot of money into recycling. We've made the decision not to force companies to clean up their own pollution. And so once people understand that it's not just a given, but it's been made to seem like a given. But if you really look into it, it's not.

The *Breathing Easy* crew took a few different actions to address the problems they were researching. First, they shot and edited a short instructional video for WEACT to show Harlem residents how to test for lead dust in their apartments. As noted above, they screened their documentary in various venues city-wide to promote community dialogue against NYCHA's environmental racism in public housing. And on a more personal level, they sought to use their documentary to pressure NYCHA to eradicate the mold from Raelene's family's home fully. These experiences not only motivated the students to become active but had a profound impact on Raelene and her mother. Michele (March 25, 2022) explains how she became an environmental justice activist:

The video made me change ... Initially, we were one family, in one apartment that was overrun with indoor environmental issues. After the video, we were a family seeing that now others were coming to us saying, "Oh, we have the same problem." That re-opened the activist part of me ... not only did I learn more, but I advocated. We spoke at several conferences. I in turn went to City Hall and spoke on news conferences ... and to Washington DC .... So, now I walk into a legislator's office, it's no longer a fear. And most people don't know that they have these privileges.

Reflecting back on what taking action meant for her, Raelene (March 25, 2022) said:

I felt like this fight was for my mother to take off with. 'Cause I like to fight a different way. She likes to go to these town hall meetings and send these emails....And I would make a video and show you the fact. I would go on and tell stories visually. And have that confidence to tell my story. Regardless of how hard it is to watch or how cringey it is for certain people. The story has to be told.

The collaborative process of documentary inquiry, research, and action was life-changing for many EVC students involved. Michele (March 25, 2022) expanded on how transformative the experience was for her daughter.

Speaking from a parent's perspective, just watching the opening and video diary brings me to tears every time I see it. EVC made a huge impact and change on Raelene. It made a change in her education, her outlook, and her future. And even though the issue regarding *Breathing Easy* still is not over, it pulled the fighter out of her. And it gave her opportunities. She went back to school. She thought she was just going to get a Graduate Equivalency Diploma (GED). Instead, she went on to get an Associates and Bachelors degrees.

Climate and Environmental Justice over the Long Haul: Carrying the Work Forward

The next generation of students is now carrying the work forward. Environmental youth movements are calling for urgent action in response to the growing global climate catastrophes. Yet, the overwhelming scale of these problems, as well as the normalization of them, has also caused many to slip into passivity, cynicism, and hopelessness. The challenge then for educators is to meet students where they are and teach them to use media to investigate how it impacts their lives in their cities, towns, and neighborhoods and make their voices heard with a sense of hope for climate and environmental justice.

This earlier generation of EVC students created media exemplars for learning, researching, and chronicling environmental citizen activism in urban communities. Animated by students' questions and community voices, their projects connected systemic environmental degradation to students' lived experiences. They started small, questioned the familiar and the given, forged intergenerational community partnerships, and used participatory research, humor, and creativity to call youth and adult audiences to action. For example, they created video stories about the early recycling centers and Hudson Riverkeeper's legal action in the late 1980s to youth and adult activists' calls for environmental justice and healthy communities in the South Bronx and Harlem in the 2010s. Doing so opened their eyes to the decades of hazardous housing and unchecked disposal of waste and toxic chemicals in our land, water, and air that caused the poorest communities to disproportionately suffer from government and corporate abuse and neglect.

These documentaries and lessons learned while creating them still speak to teachers and students today who are engaging in YPAR work at the intersection of the climate crisis and environmental justice. Teachers can leverage community assets and students' considerable creative skills in media production and social media networking for research, documentation, storytelling, and organizing to create instructional public health videos, action research projects with environmental justice organizations, or other media projects.

Key to the student-centered YPAR approach is engaging students in their community outside school as a vibrant space for critical research, problem-solving, and filmmaking. Although administrative and parental permissions are usually required, and transportation is often needed, the rich opportunities for authentic student learning and media creation make it well worth the trouble.

Climate projects will vary depending on students' grade level, class and project group size, technology access, frequency and duration of classes or after-school clubs, administrative support, teacher experience, and other factors. Whether the YPAR media projects are introductory or in-depth, teachers will need to scaffold skill-building support for students so they can most effectively use media to convey their messages and make their voices heard.

By facilitating a wide range of possible activities embedded in the production process, students can learn to identify a key climate justice problem that they want to investigate and change; interview peers and community members about their experiences with the climate problem and possible solutions; map their knowledge, new questions, and who best can answer them; partner with community environmental advocates, artists, or public scientists for further video documentation, research, and action; create music, animation, spoken word, or other elements of creative expression; review video footage and create an edit plan; create an action plan using their video to educate audiences and inspire change; edit and present their video to spark remote and in-person dialogue and climate activism in school, community and beyond (Goodman, 2018, p 139).

The following are some possible lines of inquiry and sources that may serve as launching pads to prompt student discussion and investigation. Students can then create videos on how extreme heat waves, hurricanes, floods, and toxic air pollution disproportionately impact the health and well-being of underserved communities of color—and what can be done to improve these conditions:

- Aging cities like New York are hit with the scorching heat that impacts all residents, but Black New Yorkers are twice as likely to die from heat as white residents are (Barnard et al., 2022). Students might document the urban tree planting and green space initiatives to support heat-vulnerable low-income communities.
- Intolerably hot classroom temperatures disproportionately inhibit learning for Black and Latinx students who are less likely to have air conditioning in their homes or schools. By 2025, an estimated one in four public schools nationwide will need to install or upgrade air conditioning systems. Students might research this problem and what can be done as urban schools are increasingly sending students home early for "heat days" (Meckler & Phillips, 2022).

- Smoke exposure from forest fires and hotter, longer warm seasons lead to an increase in air pollutants, pollen, and other allergens that are especially harmful to people suffering from asthma and other respiratory illnesses (Centers for Disease Control and Prevention, n.d.). Students might research air quality and disparities in respiratory illness in their community.
- Mold outbreaks in low-income housing followed Hurricane Katrina in New Orleans and Hurricane Sandy in New York (Pike, 2020). Students might research how the respiratory health of low-income housing residents can be safeguarded as risks grow of hurricanes and floods, and crumbling infrastructure.
- Poor residents who live in basement apartments are vulnerable to flash floods, as was the case when 11 New Yorkers died when stormwater filled their basement apartments after Hurricane Ida (Barnard et al., 2022). Students might investigate housing codes and environmental safety measures proposed to prevent more deaths.
- An 85-mile stretch between Baton Rouge to New Orleans, Louisiana (dubbed "Cancer Alley") has the densest concentration of petrochemical plants in the country, located near mostly Black and poor towns, with a cancer risk rate 50 times the national average. Students might research community efforts to document and change similar toxic housing patterns for communities of color (Cho, 2020; Laughland, 2022).

These are just a few possible climate and environmental justice problems for which students can use media to investigate and propose solutions. For example, at the time of this writing, extreme floods and disinvestment in broken water systems have left more than 150,000 predominantly low-income Black residents of Jackson, Mississippi without drinking water. In our climate-changed world, student voices advocating for sustainable and just environmental action are needed now more than ever.

Acknowledgment

The author thanks Raelene Holmes-Andrews, Michele Holmes, Christine Mendoza, and David Murdock for their generous time and assistance.

References

Almassi, B. (2021). *Reparative environmental justice in a world of wounds*. Rowman and Littlefield.

Barnard, A., Kilgannon, C., Hughes, J., Goldberg, E., & Mei-Ling, S. (2022, May 28). It's going to be a hot summer. It will be hotter if you're not rich. *The New York Times*. https://www.nytimes.com/2022/05/28/nyregion/heat-waves-climate-change-inequality.html

Cammarota, J., & Fine, M. (Eds.). (2008). *Revolutionizing education: Youth participatory action research in motion*. Routledge.

Caraballo, L. (2022). YPAR as figured world: Co-authoring identities, literacies, and activism. In Beach, R. (Ed.), *Drawing on students' worlds in the ELA classroom: Toward critical engagement and deep learning* (pp. 119–141). Routledge.

Centers for Disease Control and Prevention. (n.d.). Climate change decreases the quality of the air we breathe. https://www.cdc.gov/climateandhealth/pubs/air-quality-final_508.pdf

Chavis, B. (1993). Foreword in confronting environmental racism: Voices from the grassroots. South End Press. 3. https://www.washingtonpost.com/climate-environment/interactive/2021/environmental-justice-race/

Cho, R. (2020, September 22). Why climate change is an environmental justice issue. State of the Planet. Columbia Climate School, Columbia University. https://news.climate.columbia.edu/2020/09/22/climate-change-environmental-justic

Educational Video Center (Producer). (1989). New York City and the Hudson River: Downstream and up the creek. [Video file] EVC. https://www.evc.org/product-page/new-york-city-the-hudson-river

Educational Video Center (Producer). (1990). Trash thy neighbor [Video file]. EVC. https://www.evc.org/product-page/trash-thy-neighbor

Educational Video Center (Producer). (2007). Shame on you! That can be reused! [Video file]. EVC. https://www.evc.org/product-page/shame-on-you-that-can-be-reused

Educational Video Center (Producer). (2013). Breathing easy: Environmental hazards in public housing [Video file]. EVC. https://vimeo.com/66192997

Fears, D., & Dennis, B. (2021, April 6). This is environmental racism: How a protest in a North Carolina farming town sparked a national movement. *The Washington Post*. https://www.washingtonpost.com/climate-environment/interactive/2021/environmental-justice-race/

Fine, M., & Torre, M. E. (2019). Critical participatory action research: A feminist project for validity and solidarity. *Psychology of Women Quarterly*, *43*(4), 433–444.

Gammon, K. (2022, May 5). US is recycling just 5% of its waste, studies show. *The Guardian*. https://www.theguardian.com/us-news/2022/may/04/us-recycling-plastic-waste

Goodman, S. (2018). *It's not about grit: Trauma, inequity, and the power of transformative teaching*. Teachers College Press.

Laughland, O. (2022, April 14). EPA opens civil rights investigations over pollution in Cancer Alley. *The Guardian*. https://www.theguardian.com/us-news/2022/apr/14/cancer-alley-louisiana-civil-rights-investigations-epa-pollution

Lippard, L. R. (2016, October 28). New York comes clean: The controversial story of the Fresh Kills dumpsite. *The Guardian*. https://www.theguardian.com/cities/2016/oct/28/new-york-comes-clean-fresh-kills-staten-island-notorious-dumpsite

Meckler, L., & Phillips, A. (2022, June 4). Climate change is forcing schools to close early for "heat days." *The Washington Post*. https://www.washingtonpost.com/education/2022/06/04/school-heat-days-climate-change/

Moll, L. C., Amanti, C., Neff, D., & Gonzales, N. (1992). Funds of knowledge for teaching: Using a qualitative approach to connect to homes and classrooms. *Theory into Practice*, *32*(2), 132–141.

New York City Department of Parks and Recreation. (n.d.). Freshkills Park. https://www.nycgovparks.org/park-features/freshkills-park/about-the-site

Petrone, R., Mirra, N., Goodman, S., & Garcia, A. (2022). Youth civic participation and activism (youth participatory action research). In J. Z. Pandya, R. A. Mora, J. Alford, N. A. Golden, & R. S. deRoock (Eds.), *The handbook of critical literacies* (pp. 50–60). Routledge.

Pike, L. (2020, April 1). Rising sea levels leave public housing residents struggling with mold. *Popular Science.* https://www.popsci.com/story/environment/public-housing-mold-climate-change/

Riverkeeper. (2009). Hudson River pcbs. https://www.riverkeeper.org/campaigns/stop-polluters/pcbs/

World Economic Forum. (2020). What is environmental racism and how can we fight it? https://www.weforum.org/agenda/2020/07/what-is-environmental-racism-pollution-covid-systemic

11

RESILIENT BY YOUTH ENGAGEMENT

The Alameda Creek Atlas

N. Claire Napawan, Brett L. Snyder,
and Beth Ferguson

Introduction

Climate change is often conceptualized as an impending ecological disaster. However, vulnerability to its impacts is more accurately defined by social factors that present daily challenges to communities. Moreover, the predicted impacts that will exacerbate vulnerability are often difficult to comprehend or easily ignored, given the global scale, extended time frames (Gilbert, 2006), and predominantly negative imagery associated with climate change (Wang et al., 2018). Collectively, this often results in the "distancing" of everyday habits and decision-making from the larger issue of climate crisis; or as Wang et al. note: "there's an absence of human stories—those that show ordinary and relatable humans engaging with the issue of climate change" (Wang et al., 2018, p. 3). Building resilience to these impacts will require engaging communities in understanding their own vulnerabilities and co-designing opportunities for adaptation (Simpson et al., 2019). It will also require communicating projected impacts to communities to encourage acceptance, dialogue, and action (Corner et al., 2015). As such, there is a need for new tools for communicating climate impacts, defining vulnerability indicators, and engaging diverse publics in planning for future resilience. This will require collaboration between designers, decision-makers, and community members and a robust conversation about the full socio-ecological dimensions of climate change through new and unique engagement strategies.

In this chapter, the authors argue for place-based participatory design and decision-making as a youth engagement tool and an opportunity to build community resilience to the impacts of climate change (Paschen & Ison, 2013).

DOI: 10.4324/9781003335276-15

The authors discuss the current challenges to and existing models of successful youth engagement with climate adaptation planning. Finally, the site-specific climate engagement project, *Alameda Creek Atlas*, is detailed as a case study for employing such a methodology, and preliminary results are discussed.

Background

Community resilience, based on ecological definitions of resilience, is the ability of a community to respond to change and disruption, while still maintaining its general function, structures, form, and identity (Allen & Bryant, 2011; Amundsen, 2012). In addition, resilience focuses on questions of the qualitative characteristics of a system, its strengths, long-term viability, and ability to learn and adapt (Allen & Bryant, 2011; Vale, 2014). Specifically, the ability to learn, adapt, and manage change becomes an important aspect of identifying and understanding climate resilience within a community (Folke, 2006; Magis, 2010; Ross & Berkes, 2014; Tyler & Moench, 2012). Thus, community resilience to climate impacts requires "communication, social equity, and participation to facilitate transformative learning processes" (Paschen & Ison, 2013, p. 1084); it also requires meaningful climate engagement.

True and meaningful engagement goes beyond mere literacy or awareness of risks; instead, climate engagement helps foster awareness and concern that promotes changes in opinion and actions, often resulting in behavior change and/or political advocacy (O'Neill & Nicholson-Cole, 2009). It is growing as a mitigation and adaptation strategy in North American cities (Ebi & Semenza, 2008), with notable examples in the more vulnerable communities of the Northwest Territories of Canada (Armitage, 2005; Cohen, 1997), Florida (Frazier et al., 2010), and California (Garzon et al., 2015; Moser & Ekstrom, 2011). These engagement strategies integrate planners, decision-makers, and stakeholders in effectively addressing the challenges associated with local and regional climate impacts, linking social and ecological factors that contribute to vulnerability or resilience within a community.

In addition, these engagement efforts give voice to those most vulnerable to climate change (Ross & Berkes, 2014). Vulnerability to climate impacts is inequitably distributed across communities throughout North America (and beyond), disproportionately impacting communities of color as well as immigrant and low-income communities (Huq & Reid, 2007; Reid et al., 2009). Though young people are typically limited in their ability to participate in the political processes surrounding climate change, those same youth are the ones that will inherit the long-term impacts of climate change (Corner et al., 2015). Climate engagement enables opportunities to address this inequity by empowering youth and other vulnerable populations to understand impacts better, plan for adaptation, and advocate for resilience.

Current barriers to meaningful climate engagement include the psychological distancing associated with the long time frames of projected impacts and the physical distance between many Americans. It is also the case that the sites of some of the more catastrophic climate-related events are often too great to make the issues feel relevant (Gilbert, 2006). Additionally, fear tactics that characterize climate change as "a terrible, immense, and apocalyptic problem" (Lorenzoni et al., 2007, p. 447) are successful at capturing people's attention and often utilized by popular media as a result. However, such an approach also leaves people feeling hopeless and unable to see their relationship to the issue (Lorenzoni et al., 2007; O'Neill & Nicholson-Cole, 2009). In contrast, communication approaches that "take account of individuals' personal points of reference (e.g., based on an understanding and appreciation of their values, attitudes, beliefs, local environment, and experiences) are more likely to meaningfully engage individuals with climate change" (O'Neill & Nicholson-Cole, 2009, p. 375).

Thus, approaches that focus on more immediate or near-term impacts, address local and regional scales, and engage with existing concerns and interests are the keys to meaningfully connecting community members with climate issues and planning strategies for adaptation and resilience. This variation across different contexts is particularly true for engaging young people with climate action and advocacy. Research suggests that people of color—including Hispanics/Latinos, African-Americans, and other non-White racial/ethnic groups—within the United States express greater concern than Whites about climate action (Ballew et al., 2020). This disparity is likely a result of the disproportionate negative climate impacts on communities of color.

Case Study: The Alameda Creek Atlas

The selected site for this chapter's inquiry is the lower Alameda Creek Watershed, located in Alameda County in the East Bay of the San Francisco Bay Area. This watershed is the largest local watershed in the region, with the potential to transport much-needed sediment to adapt to predicted rising sea levels in the bay. A four-foot (1.4 meters) rise in sea levels, as modestly predicted by the year 2100, would place critical energy, drinking water, waste management, transportation infrastructure, and over 270,000 residents at risk throughout the Bay Area (Jerrett, et al., 2012).

The Alameda Creek provides one of the best opportunities for testing strategies to adapt wetlands and marshes to rising sea levels through sediment transport, thus mitigating risks such as neighborhood flooding. Increased development within the creek's floodplain throughout the twentieth century necessitated the construction of upland dams and a flood control channel, dramatically decreasing the conveyance of sediment to the Bay while preventing routine flooding (San Francisco Estuary Institute, 2013). The Alameda Creek Watershed has the added potential to support steelhead trout populations by providing a route to upland spawning, should the impediments to sediment transport be addressed. Steelhead Trout has been listed as an endangered species since 1997. The Alameda Creek could provide an ideal habitat for trout migration and spawning if major barriers were removed from the creek. Reintroducing this species would increase nutrient density and complexity within the food web of the watershed, supporting many other plant and animal species in the region (Ibid).

However, to support upstream steelhead and sediment migration downstream, dramatic changes in the creek's physical design and management strategy are required to allow unimpeded up- and downstream flows and effectively "unlock" the Alameda Creek. (See Figure 11.1 for an illustration of the Alameda Creek Watershed and its potential for building resilience to climate change.)

In addition to these environmental challenges with the creek, community members within the watershed have little to no physical or visual access to the creek. Throughout the lower watershed, the creek is channelized into a concrete canal that traverses several highly engineered dams. As a result, the creek feels more like an engineered infrastructure than a social or environmental resource.

The project's site selection and preliminary research and design arose as a result of the Resilient by Design Bay Area Challenge, "a year-long collaborative design challenge bringing together local residents, public officials and local, national and international experts to develop innovative community-based

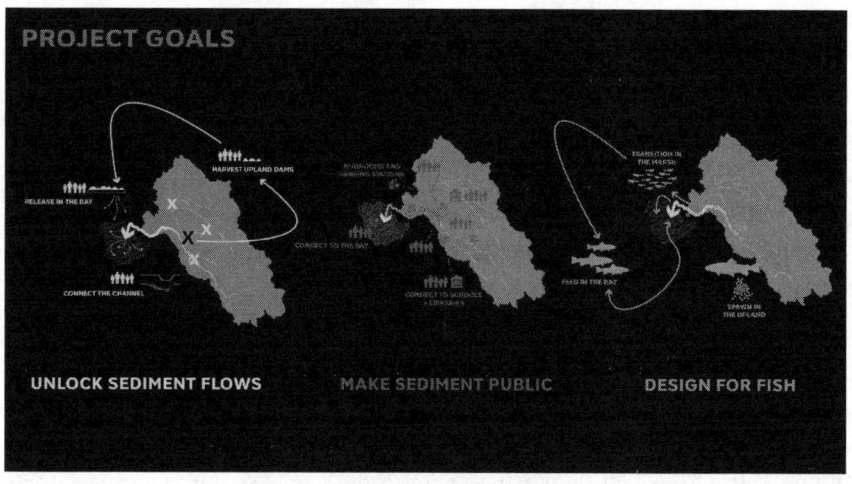

FIGURE 11.1   The Alameda Creek Watershed and the Public Sediment Team Design Goals

solutions that will strengthen our region's resilience to sea level rise, severe storms, flooding and earthquakes," that began in September 2017 and concluded in May 2018 (Resilient by Design, 2017, n.p.).

The selected team for the investigation of the Alameda Creek, Public Sediment, included

- environmental designers from Scape Studios, a landscape architecture firm based in New York City;
- Dredge Research Collaborative, a consortium of sediment-focused environmental design academics;
- Arcadis, nationally based coastal engineers;
- TS Studio, a San Francisco-based urban design firm;
- Architectural Ecologies Lab, a research and design collaborative; and
- faculty from the University of California, Davis' Departments of Design and Human Ecology, with expertise in community-based climate resilience and participatory design methodologies.

The design and planning process of this climate adaptation project required collaboration with an extensive list of regional stakeholders with jurisdiction in the watershed: Alameda County Flood Control, Alameda County Water District, Alameda County Sustainability Office, Alameda Creek Watershed Forum, Zone 7, San Francisco Public Utilities Commission, City of Fremont's Office of Sustainability, and the Alameda Creek Alliance. Also included in the lower watershed are some of the most diverse communities in the cities of Fremont, Union City, and Newark; they represent some of the most vulnerable communities to the projected impacts of climate change (including communities of color, youth, and/or low socio-economic status) (Cooley et al., 2012). Through the collaboration of designers and stakeholders, a range of phased strategies were proposed to move sediment downstream within the watershed effectively, support trout spawning upstream, and increase community access and awareness of the Alameda Creek (see Figure 11.2).

This chapter focuses specifically on the youth engagement methodology employed within the communities of the lower watershed that occurred throughout the design and planning process, and the impacts of this engagement on the project's preliminary results. Community engagement represented a critical component of the design process. The Resilient by Design's articulated grounding principles included the following statement: "Develop equitable planning and development practices where community members are true collaborators and participate as equal partners at every level of design formation" (Resilient by Design, 2016, n.p., *resilientbayarea.org*). This emphasis should come as no surprise, given the long history of successful environmental justice

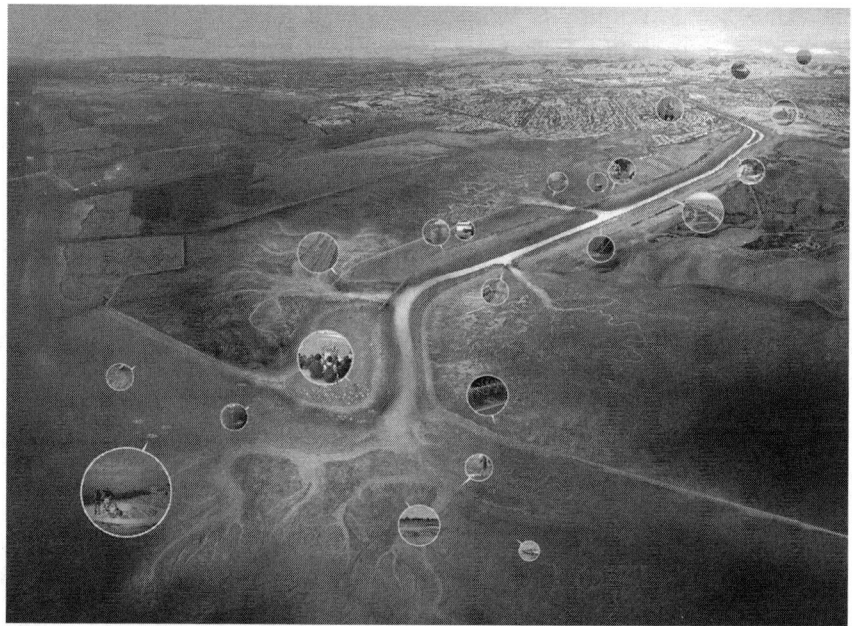

FIGURE 11.2   Unlocking Alameda Creek Design and Management Strategy, Proposed by Public Sediment Team

planning efforts in the Bay Area rooted in community engagement and action; these efforts rarely result from top-down municipal planning.

Furthermore, as mentioned prior, engagement is recognized as an increasingly valuable tool in building community resilience by encouraging attitude and behavior changes within a public constituency. The participation of youth and youth of color in these engagement efforts was particularly critical to the process. As mentioned prior, youth are particularly disproportionately more vulnerable to climate impacts than the general population and with less capacity to engage in political discourse (Napawan et al., 2017).

Methods

The project utilized three formats for community outreach and engagement: (1) public presentations, (2) integrated community events, and (3) community workshops. Public presentations were coordinated with local agencies and provided opportunities to disseminate the team's design process to regionally based organizations. Attendance was high for these events, but interaction with attendees was limited to the question/answer period following the presentation. Integrated community events allowed the engagement efforts to tap into existing local programs and events, such as community festivals and other events. In addition, events were selected based on their potential alignment with climate-related issues, such as Earth Day celebrations, and it enabled reaching an audience already interested in related matters.

Because local municipalities or non-governmental organizations often organized these events throughout the watershed, they represented the engagement process' highest attendance numbers and allowed the opportunity to engage participants more meaningfully in the climate-related issues affecting the watershed. In addition, play-based tools, such as a peg model for active mapping, stamping and drawing art exercises, and table-top sandboxes for sediment co-design made the material relevant to the youngest participants. Social media scavenger hunts encouraged digitally engaged adolescents and young adults to find significant ways to contribute to their conceptualizations of the watershed and its risks.

Lastly, community workshops provided greater collaboration with partners and focused time with participants. While these final engagement events were limited in attendance, ranging from 8 to 30 community members, they allowed for more in-depth engagement and detailed and nuanced feedback from participants. These workshops were planned with community partners, such as schools, youth centers, senior centers, and other local non-profits, and were designed to meet the mutual goals of the design team and the organizing entity. Detailed narratives of the Alameda Creek's past provided by local elders, future visions mapped on-site by local youth, and new architectural forms for protecting spawning trout represent some of these workshops' broad range of outcomes. (See Figure 11.3 for location, dates, and attendance numbers for the three engagement strategies that were utilized from January 2018 to May 2018.)

Throughout the range of formats and audiences, key concepts about the watershed were presented to the community by way of an Alameda Creek Appendix distributed at each event. The appendix served as an activity book for a broad range of ages. The appendix defined key terms such as "watershed" and "sediment," and their relationships to community climate risks were described in English, Spanish, and Mandarin, helping to build local climate literacy and awareness. Participants were then asked to provide their interpretations of their community through the Alameda Creek Field Notes, sharing their connection to the landscape, stories of floods, commute routes, community concerns, and messages for stewardship. (See Figure 11.4 for examples of prompts from the Atlas.)

These tools and the community feedback gathered through them collectively form the Alameda Creek Atlas. It exposes a watershed disconnected from the surrounding community, highly engineered yet threatened by the very development its construction sought to encourage. It also reveals a diverse and

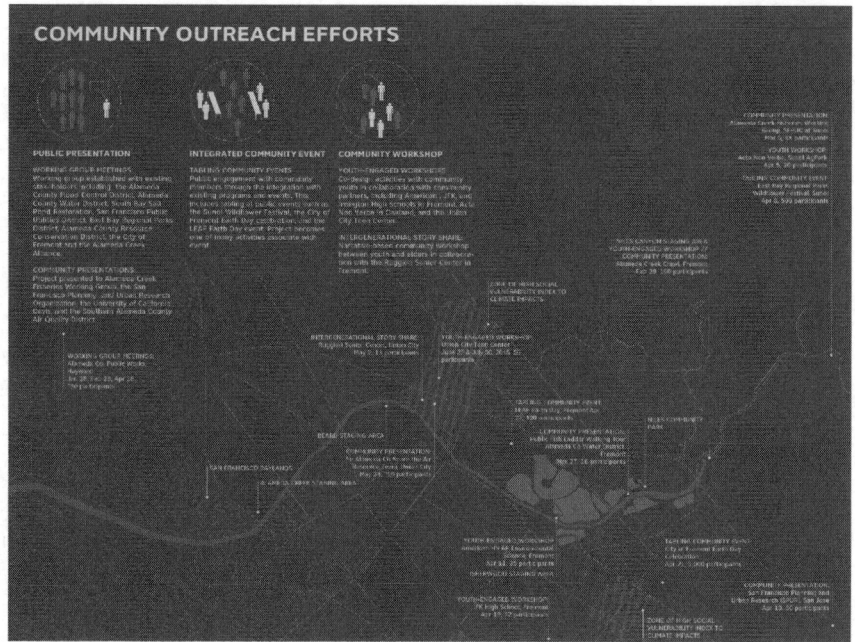

FIGURE 11.3   Outreach and Engagement Strategies and Locations

vibrant community of people eager to engage in stewardship and advocate for climate resilience.

Youth-based Community Workshops

While all methods and formats described above were designed for multi- and inter-generational engagement, a particular focus was given to youth and youth of color engagement. The most successful youth-engaged workshops included the Alameda Creek Crawl, the City of Fremont Earth Day festivals, and the Union City Teen Center workshops.

Alameda Creek Crawl

Alameda Creek Crawl was a public event held on Saturday, February 24, 2018, from 10:00 am to 12:00 pm at the Niles Canyon Staging Area, one of only a few locations where the public can access the creek within the lower watershed. The event was broadly publicized in local news outlets, event boards, and the SF Resilient by Design networks, including the Alameda Creek Conservatory and other regional stakeholders. Over 100 participants joined Creekside tours

FIGURE 11.4 Prompts from the Alameda Creek Atlas and Assorted Responses from Youth Participants

led by ecologists, urban planners, and landscape architects, communicating potential climate risks.

A range of activities were designed to interest youth during the event: a watershed painting activity that utilized crumpled paper, washable markers, and a spray bottle; native clay for modeling into potential creek channel forms; sidewalk chalk distributed throughout the tours to encourage highlighting objects in the landscape of interest or concern; and finally, a digital scavenger hunt to solicit feedback on potential new programming alongside the creek. (See Figure 11.5 for examples of youth engagement with sidewalk chalk writing and drawing at the Creek Crawl.)

The most popular activity among adolescents and young adults integrated with social media is an activity that met youth "where they already are" (Corner et al., 2015). The digital scavenger hunt encouraged participants to find evidence of

FIGURE 11.5   Young Participants at the Alameda Creek Crawl Use Chalk to Write and Draw Their Understanding of the Watershed

different conditions of resilience and vulnerability surrounding the creek; this included elements of human use as well as environmental characteristics. The exercise prompted participants to observe more closely the watershed and built environment surrounding them, to imagine alternative futures, and to share all these perspectives with their networks on social media. The activity had longevity beyond the event itself: youth continued to post and comment contributions using the tag #AlamedaCreekAtlas, thus allowing the opportunity to reflect following the event and integrate their knowledge with other aspects of their everyday lives (Napawan et al., 2017). (See Figure 11.6 for examples of youth participation in the digital scavenger hunt.)

For younger participants, including those between the ages of 3 and 12, the most popular activities of the Alameda Creek Crawl were those that provided more tactile experiences with the watershed. This included the watershed painting activity, modeling with clay, and drawing and writing with sidewalk chalk. The watershed painting activity involved the crumpling of a sheet of heavy craft paper into a fictional landform and using blue washable markers to trace the logical water networks within the crevices and "low-lying terrain" of the crumpled paper. Participants were then asked to use a red washable marker to indicate the "best" places for development that would cause the minimum amount of impact to the "watershed." Finally, participants simulated rainfall

FIGURE 11.6   Examples of Instagram Posts from Youth Participants Engaging with the Alameda Creek Digital Scavenger Hunt

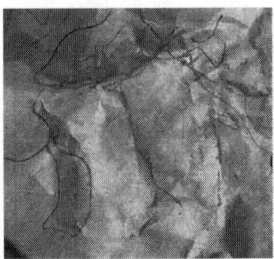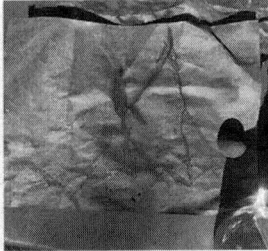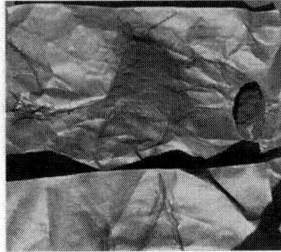

FIGURE 11.7 Youth Participants' Paper "Watershed Terrains" to Explore the Impact of Watersheds on Development and Vice Versa

on their landscape using a spray water bottle and were asked to evaluate the impact of the development (red ink seepage) on the watershed (blue ink seepage) and vice versa. In every case, the youth participants learned that there was no possible way to limit the impact of watersheds on development or development on watersheds. (See Figure 11.7 for examples of paper watershed activity completed by youth participants.)

Earth Day Festivals

Within the lower watershed, there is a range of festivals in April to recognize the significance of Earth Day. These events represent an opportunity to connect with community members that live in Fremont, Union City, and Newalk who are already interested in environmentally conscious issues, and thus an opportune moment to connect them to issues surrounding the Alameda Creek and climate impacts. Pre-planned festivals were held on Saturday, April 21, 2018, at the City of Fremont Convention Center and Sunday, April 22, 2018, at the headquarters of the local non-profit Local Ecology and Agriculture Fremont (LEAF). The City of Fremont's event was attended by thousands of residents and stakeholders and the LEAF event had hundreds in attendance throughout the day. At both Earth Day festivals, families with school-aged children were the most common attendees. The most popular workshop activity at these events (for youth and adults alike!) was an interactive creek bed "sandbox." The sandboxes were crafted as relief maps of several problematic locations of the Alameda Creek where sediment currently gets trapped. Using kinetic sand, youth participants could better understand the impacts of the creek channel on sediment flow; most importantly, they could manipulate the sand to explore different shapes and forms that might improve function. 3D-printed objects were also included in the sandboxes to indicate different constituents and programs that might utilize the creek. (See Figure 11.8 for the creek channel sandboxes in use at the Earth Day festivals.)

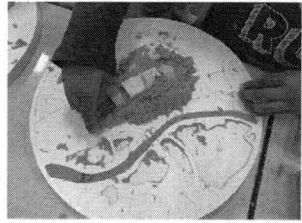

FIGURE 11.8   Sandboxes Shaped like the Alameda Creek Channel Allow Youth to Better Understand and Manipulate Sediment within the Watershed

Another popular activity among youth participants at the Earth Day festivities was a participatory mapping exercise that utilized a topographic model of the lower Alameda Creek Watershed. The model was a scaled representation of community members' landscape, but with limited distinguishing features. Round pegs were placed on a grid throughout the model and long brightly colored rubber bands were stretched across the pegs to indicate existing flows through the topographic landscape (see Figure 11.9). The model provided a new perspective to the landscape of the watershed in which youth participants inhabited; it also exposed how vulnerable the lower watershed truly is, with the majority of development, transportation, and other essential infrastructure located within the Alameda Creek's floodplain.

Union City Teen Center Workshops

Workshops at the Union City Teen Center were held on Monday, June 25, and July 30, 2018. Participants in the workshop included middle-school students enrolled in summer day camp who ranged between 11 and 14 years of age. The Teen Center is located less than a half-mile from the Alameda Creek and in a designated high social vulnerability index for climate impacts zone (San Francisco Bay Conservation and Development Commission [SF BCDC], 2020).

FIGURE 11.9   A Topographic Model with Pegs and Rubber Bands Allows Young People to Explore Flows of Water, People, Sediment, and Wildlife within the Context of Their Landscape

FIGURE 11.10   Union City Teen Center Workshop Participants at the Alameda Creek

The first workshop introduced climate impacts within the watershed and allowed facilitators to hear student perspectives on climate impacts and the creek. Social media games with prizes encouraged students to scroll through their feeds for evidence of climate vulnerability or resilience in their own lives. Much like prior digital scavenger hunt activities, this allowed students to engage with the social media networks that are already significant to them. Unlike on-site games, this also allowed students to recognize that their daily lives might already have a relationship to climate impacts. By scrolling through existing content, as opposed to generating new content, the exercise pushes youth to consider what places, activities, and people might be already impacted by climate change.

The second workshop was held on-site at the Niles Canyon Staging Area; although the site is within walking distance to the Teen Center, students had to be bused to the site due to the lack of public access to the creek from the North. While on-site, students utilized models, large-scale drawings, and/or survey flags to envision resilient futures for the Creek. A scaled physical model of the site allowed participants to better understand relationships between development, watershed, and climate resiliency; bringing the model to the site allowed youth to translate to a real place and fully scale the implications of climate change in their built environments. Students also explored alternative scenarios for future development and resilience planning through large-scale drawings and adding structures to the scaled model. By translating those ideas onto the physical landscape at full scale using survey flags, the translation to real-world implications of their design ideas became more accessible. (See Figure 11.10 for images of youth at the on-site climate workshop with various partial- and full-scaled tools.)

Conclusion

Aside from integrating community youth within the watershed with the Resilient by Design's adaptation design and planning process, the Alameda Creek

Atlas became a community-building tool itself: highlighting existing community connections and encouraging new ones. The best indicator of resilience to environmental disasters to date is community connections: as demonstrated by the #OccupySandy movement in New York and New Jersey following Superstorm Sandy in 2012 (U.S. Department of Homeland Security, 2013); and by recovery efforts in Japan following the 2011 earthquake and tsunami: "Standard advice about preparing for disasters focuses on building shelters and stockpiling things like food, water, and batteries. But, resilience—the ability to recover from shocks, including natural disasters—comes from our connections to others, and not from physical infrastructure or disaster kits" (Aldrich, 2017, p. 1).

Current efforts at understanding social vulnerability to climate change within the region often do not include opportunities for the full public—including youth—to inform the selected indicators list. For example, social vulnerability, as defined in the San Francisco Bay Area by the Pacific Institute in 2012, neglected some indicators that were identified as community concerns, including transportation equity and proximity to polluting industries (SF BCDC, 2020). In addition, indicators of climate vulnerability can quickly become outdated, as was demonstrated by the March 2018 release of subsidence data in the San Francisco Bay Area—six months *after* the start of the design competition. This information revealed the need to revise models for evaluating inundation risk from flooding and sea level rise (Griggs, 2018).

In response to these challenges, we need adaptive models of information sharing, a better merger of social and ecological conceptualizations of climate changes, and most importantly to employ community-engaged approaches to allow climate change to be perceptible at the local level, *to be known*. The Alameda Creek Atlas provides a model for the place-based understanding of community risk, change, and resilience by youth engagement. When utilized within the classroom, these tools bring a greater connection between climate change as an academic subject and students' lived experiences. Typical climate change curriculum is taught as an extension of environmental or atmospheric sciences and as a result, the implications of the curriculum on students' everyday lives are often not clear (Dupigny-Giroux, 2010).

The tools described in this chapter represent experiential learning activities and a merger of social science and humanities perspectives with environmental sciences that bring a greater connection between youth interests and climate impacts. In addition, the examples in this chapter use localized contexts for exploring climate impacts in contrast to the conventional globally scaled approaches, allowing content surrounding climate change to feel more relevant to students' lives (Kiers et al., 2020). Finally, and most significantly, these youth engagement techniques provided an opportunity for young people to imagine new and alternative futures, building in them the capacity to achieve greater climate resilience for themselves and their communities.

References

Aldrich, D. (2017). Recovering from disasters: Social networks matter more than bottled water and batteries. *The Conversation.* Retrieved March 23, 2018, from https://theconversation.com/how-social-ties-make-us-resilient-to-trauma-78223

Allen, P., & Bryant, M. (2011). Resilience as a framework for urbanism and recovery. *Journal of Landscape Architecture, 6*(2), 34–45.

Amundsen, H. (2012). Illusions of resilience? An analysis of community responses to change in northern Norway. *Ecology and Society, 17*(4), 46–59.

Armitage, D. R. (2005). Collaborative environmental assessment in the Northwest Territories, Canada. *Environmental Impact Assessment Review, 25*(3), 81–101.

Ballew, M., Maibach, E., Kotcher, J., Bergquist, P., Rosenthal, S., Marlon, J., and Leiserowitz, A. (2020). *Which racial/ethnic groups care most about climate change? Yale University and George Mason University.* New Haven, CT: Yale Program on Climate Change Communication.

Cohen, S. J. (1997). Scientist-stakeholder collaboration in integrated assessment of climate change: Lessons from a case study of Northwest Canada. *Environmental Model Assessment, 2*(4), 281–293.

Cooley, H., Moore, E., Heberger, M., & Allen, L. (2012). Social vulnerability to climate change in California. Retrieved March 11, 2018, from http://pacinst.org/

Corner, A., Roberts, O., Chiari, S., Voller, S., Mayrhuber, E., Mandl, S., & Monson, K. (2015). How do young people engage with climate change? The role of knowledge, values, message framing, and trusted communicators. *WIREs Climate Change, 6,* 523–534.

Dupigny-Giroux, L.-A. L. (2010). Exploring the challenges of climate science literacy: Lessons from students, teachers and lifelong learners. *Geography Compass, 4,* 1203–1217. doi:10.1111/j.1749-8198.2010.00368.x

Ebi, K. L., & Semenza, J. C. (2008). Community-based adaptation to the health impacts of climate change. *American Journal of Preventive Medicine, 33*(5), 501–507.

Folke, C. (2006). Resilience: The emergence of a perspective for social-ecological systems analyses. *Global Environmental Change, 16,* 253–267.

Frazier, T. G., Wood, N., & Yarnal, B. (2010). Stakeholder perspectives on land-use strategies for adapting to climate-change-enhanced coastal hazards: Sarasota, Florida. *Applied Geography, 30*(4), 506–517.

Garzon, C., Cooley, H., Heberger, M., Moore, E., Allen, L., Matalon, E. Doty, A., & The Oakland Climate Action Coalition. (2015). Community-based climate adaptation planning: Case study of Oakland, California. A report prepared for the California Energy Commission by the Pacific Institute. Retrieved October 10, 2015, from http://www.energy.ca.gov/2012publications/CEC-500-2012-038/CEC-500-2012-038.pdf

Gilbert, D. (2006, July 2). If only gay sex caused global warming. *The Los Angeles Times,* p. 14.

Griggs, T. (2018, March 7). More of the Bay could be underwater in 2100 than previously thought. *The New York Times.* https://www.nytimes.com/interactive/2018/03/07/climate/san-francisco-sinking-land-flooding-climate-change.html

Huq, S., & Reid, H. (2007). Community-based adaptation: A vital approach to the threat climate change poses to the poor. IIED Briefing Paper, International Institute for Environment and Development, London.

Jerrett, M., Su, J. G., Reid, C. E., Jesdale, B., Ortega, A. M., Hinojosa, S. B., Shonkoff, E. S., & Morello-Frosch, R. (2012). *Mapping climate change exposures, vulnerabilities, and adaptation to public health risks in the San Francisco Bay and Fresno Regions.* California Energy Commission, University of California, Berkeley. Publication number: CEC-500-2012-041.

Kiers, A. H., de la Peña, D., & Napawan, N. C. (2020). Future directions—Engaged scholarship and the climate crisis. *Land, 9*(9), 304. doi:10.3390/land9090304

Lorenzoni, I., Nicholson-Cole, S. A., & Whitmarsh, L. (2007). Barriers perceived to engaging with climate change among the UK public and their policy implications. *Global Environmental Change, 17*, 445–459.

Magis, K. (2010). Community resilience: An indicator of social sustainability. *Society and Natural Resources: An International Journal, 23*(5), 401–416.

Moser, S., & Ekstrom, J. (2011). Taking ownership of climate change: Participatory adaptation planning in two local case studies from California. *Journal of Environmental Studies and Science, 1*, 63–74.

Napawan, N. C., Simpson, S., & Snyder, B. (2017). Engaging youth in climate resilience planning with social media: Lessons from #OurChangingClimate. *Journal of Urban Planning, 2*(3), 1–13.

O'Neill, S., & Nicholson-Cole, S. (2009). "Fear won't do it." Promoting positive engagement with climate change through visual and iconic representation. *Science Communication, 30*(3), 355–379.

Paschen, J. A., & Ison, R. (2013). Narrative research in climate change adaptation—Exploring a complementary paradigm for research and governance. *Research Policy, 43*, 1083–1092.

Reid, H., Alam, M., Berger, R., Cannon, T., Huq, S., & Milligan, A. (Eds) (2009). Special issue: Community-based adaptation to climate change. Participatory Learning and Action, 51(4). doi:10.3200/ENV.51.4.22-31

Resilient by Design. (2016). Homepage. http://www.resilientbayarea.org/

Resilient by Design. (2017). Resilient by Design Bay Area Challenge. Retrieved March 3, 2018, from http://www.resilientbayarea.org/

Ross, H., & Berkes, F. (2014). Research approaches for understanding, enhancing, and monitoring community resilience. *Society and Natural Resources: An International Journal, 27*(8), 787–804.

San Francisco Bay Conservation and Development Commission. (2020). Adapting to Rising Tides: Regional Vulnerable Communities Section. Retrieved October 4, 2022, from http://www.adaptingtorisingtides.org/wp-content/uploads/2020/03/ARTBayArea_Regional_VulnerableCommunities_Final_March2020_ADA.pdf

San Francisco Estuary Institute. (2013). Alameda Creek watershed: Historical ecology study. SFEI Publication #679, Richmond, CA. Retrieved March 11, 2018, from http://www.sfei.org/projects/AlamedaCreekHE

Simpson, S., Napawan, N. C., & Snyder, B. (2019). #OurChangingClimate: Building networks of community resilience through social media and design. Geohumanities. doi:10.1080/2373566X.2019.1575761

Tyler, S., & Moench, M. (2012). A framework for urban climate resilience. *Climate and Development, 4*(4), 311–326.

United States Department of Homeland Security. (2013). *The Resilient Social Network.* Retrieved March 11, 2018, from http://us.resiliencesystem.org/department-homeland-security-resilient-social-network-occupysandy-superstormsandy

Vale, L. J. (2014). The politics of resilient cities: Whose resilience and whose city? *Building Research and Information, 42*(2), 191–201.

Wang, S., Corner, A., Chapman, D., & Markowitz, E. (2018). Public engagement with climate imagery in a changing digital landscape. *WIREs Climate Change, 9.* doi:10.1002/wcc.509

12

ELEVATING YOUNG VOICES

Media Created by Youth for Youth

Julian Arenas, Ardra Charath, and Liane Xu

An Introduction to the Magazine

The Youth Think Climate (YTC) magazine is a digital, youth-led publication dedicated to showcasing the thoughts and experiences of youth on the climate crisis through a variety of artistic mediums. With the support of Action for the Climate Emergency *acespace.org* (ACE), formerly Alliance for Climate Education, an organization that seeks to empower young climate justice activists and pass a national Climate Emergency Act, the magazine was formed to build a community within the climate movement so that youth have a space to self-express and develop the belief that they can create change. By placing an emphasis on storytelling and personal narratives, YTC brings in fresh, global perspectives and encourages youth to reflect on how the climate crisis affects them on a personal level. Many of the pieces featured in YTC have centered around a defining moment in the author's experience with climate change. Past editions have delved into various topics, including climate denialism, justice, solutions, and sustainability. Each installment displays a plethora of creative submissions, from poetry and short stories to paintings and drawings, that relate to the theme of an edition (*issuu.com/youththinkclimate*).

How It Started

The magazine was first developed in the early months of the COVID-19 pandemic in 2020 through a partnership between six high school students from Orlando, Florida and Madison, Wisconsin. At that time, ACE ran semester-long fellowships which served as cohorts for young people to learn various

DOI: 10.4324/9781003335276-16

skills such as writing personal narratives, leading discussions about climate change, and the importance of climate justice. This allowed fellows to then engage critically in the climate movement and take action within their communities. As the Spring 2020 Fellowship came to a close, Abby Ross, a youth organizer with ACE at the time, encouraged fellows to take the skills they had developed during the program and apply it to the real world. Recognizing the importance of the youth voice surrounding climate change and the power of sharing content online, the six fellows sought a way to provide a digital outlet for these voices, and thus the YTC magazine was born.

The magazine's purpose is to provide youth with an audience and community within the climate movement; however, it is meant to be read by anyone, no matter their age or involvement with climate activism. In general, we strive to present artwork from people who are 25 and under. The vast majority of our submitters are middle and high school students. Apart from youth submissions, we also include interviews we have conducted within the magazine. Typically, these have been with more experienced (and in most cases, older) climate activists with strong roots within the climate movement. The interviews focus on providing guidance, motivation, and insight to the reader based on the perspectives of the individuals we interview. After publication, our magazine is shared online through the Instagram accounts of both ACE and YTC, ACE's blog, and the Youth Action Network, a nation-wide community of thousands of young people who have signed up for SMS updates, information, and resources from ACE regarding the climate movement. Our Instagram following comprises about 50% 13–24-year-olds, 23% 25–34-year-olds, 14% 35–44-year-olds, and 13% 45+-year-olds. While our audience is composed of primarily younger people, there are a sizable number of individuals who are older.

Elevating Underrepresented Voices

YTC focuses on elevating the voices of young people, who may face pushback from family or community members who do not believe in the validity of climate change or its urgency, thus creating a need for these individuals to communicate, process, and understand their emotions, thoughts, and the perspectives of others. YTC aims to fill this gap by providing an outlet for young people to express themselves in relation to the climate crisis. Our submitters come from many geographical regions, and their discussions of climate justice and experiences related to the matter are often included in their artwork (featured later in this chapter). This leads us to believe that YTC has provided a platform for individuals within marginalized groups that often bear the brunt of climate change while hardly contributing to or exacerbating the crisis itself who are historically people of color, low-income people, inhabitants of the Global South, and queer people.

Benefits of a Digital Magazine

YTC, and the medium of digital communication as a whole, presents an entirely new form of accessibility to the reader. Anyone with access to the Internet has the ability to read the editions and submit work at no cost. As a result of the COVID-19 pandemic, there was no reasonable way to print and distribute the first edition in a safe and cost-effective manner, so YTC was published online instead. It is uploaded to Issuu.com, which hosts digital publications. This allows for the magazine to be easily sent across the world between readers and organizations. If our work was presented through a physical publication, printing and distribution costs would likely mean that it could no longer be free and would therefore be less accessible. Similarly, other forms of presentation, such as an exhibition in an art gallery, would introduce various fees, and, because it is geographically immobile, can only cater to a select audience. This would prevent YTC from supporting and uplifting the wide span of people who are featured in the magazine.

Team Structure

As of September 2022, YTC is run by a team of ten, which includes the three authors of this chapter, as well as a youth organizer who serves as a liaison between YTC and ACE. The YTC team is divided into two sub-teams. The first is the Writing/Editing Team, which is responsible for selecting submissions to be included in the magazine, managing communications, interviewing activists for the magazine, writing posts for the blog, and writing sections of the magazine. The second is the Graphic Design/Social Media Team, which handles importing accepted submissions into the design software, creating the look of the magazine, making social media posts (including submission spotlights, announcements, and other informative posts), and reaching out to other related organizations on social media for partnerships and mutual promotion. Originally, YTC was organized into four groups, though the logistical complications and overlap of various tasks led to the decision to consolidate them for Edition 3 and its following issues.

A group of about four to five work in each team, with one or two designated "team leads" that oversee the group's success, spearhead and organize projects, and help other members develop and learn new skills. Additionally, they meet with the other team leads to share updates with each other, make decisions, and direct the overarching path of the magazine. Currently, the authors of this chapter serve as team leads. Members of the two sub-teams applied through an online process in which they answered several questions and briefly met with a team lead. Many of these individuals, who are students from across the country, had submitted work to YTC in the past and were interested in participating further with the group.

The Submission Process

YTC utilizes social media to share information about the magazine, including how to submit work. Team members design several Instagram posts with information about the magazine and its theme, and individuals can find the submission form on YTC's Instagram profile. Calls for submissions are also publicized through ACE's Instagram and Youth Action Network *t.ly/_9fq*.

Submissions are selected for the magazine based on two criteria:

1 Relation to the Edition Theme: While the connection does not need to be overt, the theme (e.g., Climate Denial, Climate Justice, and Sustainability) should be mentioned, implied, or addressed in some way within the piece. As our goal is to create a cohesive collection of media with the end product of the issue, this is a key benchmark to ensure the magazine does not feel disjointed.
2 Artistic Merit: We cannot include every piece we receive, so we must ask ourselves while reading submissions:

a What perspective or voice does this bring to the edition?
b How does it relate to the broader climate movement?
c What is unique or noteworthy about this piece?

Each submission (in both the context of the piece itself and the magazine as a whole) is looked over by several editors to ensure fairness in this process. Discussions are then held to finalize the selected works and to solve any disagreements.

As we strive to present as much artwork as possible, if a certain piece does not meet the criteria, we try to work with the submitter to make sure that it can be included or make note of it for future editions in which it may reflect the theme. We do not treat the magazine as a "competition" or cap the number of accepted submissions; providing exposure and creating meaningful dialogues about the climate crisis is first and foremost our priority. One way we encourage this is by including optional spaces for submitters to write about what inspired or informed their piece and its creation and meaning. These responses are then presented alongside the artwork in the magazine and provide readers with greater context to the content. The artwork we feature also informs the organization, layout, and tone of the magazine.

Edition 1: Climate Denial

The first edition of the YTC magazine, *Climate Denial*, was released in July 2020 (Figure 12.1). It featured 29 works from eight different states across the US. The theme was inspired by an incident a YTC team member had once experienced. While volunteering at a local hospital, they were stopped in the middle of their task by another volunteer's question: "Kid, what do you enjoy?

FIGURE 12.1   "Climate Denial" Cover Image

The climate. Let me tell you why they are lying to you." The volunteer claimed that K-12 schools were using the climate justice movement as a means to control students through fear-mongering. The YTC member had no idea how to respond. The topic of climate denial was chosen with the goal of examining how others have dealt with similar situations and the experience of living in a society in which substantial climate education is not widespread. This can lead people to deny climate change and its associated ideals, hindering efforts to make sustainable change.

Some submissions featured in this edition examined the root cause of the issue of climate denial or how it was preventing sustainable progress. Similarly, many pieces were personal narratives that zeroed in on a moment or experience with climate denial. Many proposed solutions to climate denial: climate education, voting for politicians who believe in and will fight against the crisis, and international climate policy.

One notable essay was 14-year-old Svanfridur Mura's piece *Sugar*. "I like to compare climate denial to sugar," she began the essay. "I was raised to think that all sugar is bad. It tastes good, but it is the root of our failing health and I would inevitably die young if I ate it." Mura explains how a dinner table conversation resulted in her brother telling her that climate change was something kids should not have to worry about, and to stop because it was making him depressed. Mura's father agreed with her brother, leaving her feeling betrayed and isolated by her family. As she notes, this made her realize that climate denial can exist everywhere:

> Because, like sugar isn't just candy and cake, climate denial isn't just saying, "No, climate change isn't real." Sugar hides in bread, in snacks, in drinks. And climate denial can be anything from saying the above to going out and buying a gas guzzler, to telling yourself, "Climate change is an issue, but not so bad that I should stress about it instead of enjoying this day with my family." Climate denial, like most things, comes in many forms and amounts; it can be pure or diluted down so that it's barely detectable.

Reflecting on this moment later, Mura felt powerless in the climate movement. The only way she felt motivated to make change was to ignore the threat of the future. She used the power of denialism, the theme of the edition, to internalize her ability to make a difference, yet she only ate this "sugar" in slight amounts, nothing to the extent to which most politicians did.

> We must teach people that yes, climate change is real, and yes, climate change is caused by humanity, yes, we need to work to stop it, because those things are all true. Yet completely accepting the truth is also to accept that we are not in a good place. In many important ways, we are losing this battle. Accepting the entire truth of climate change is accepting the large and looming and likely possibility that we may lose, and life as we know it may come to an end. And I don't think that's a truth that many people are able to face head-on and still keep going. At least, I can't. We need to deny the possibility that that can happen. To keep going, climate activists (or anyone worried about the state of our planet) need to wake up each morning and say to themselves, "Yes, climate change is real, and we will definitely stop it." Is denying that possibility a kind of climate denialism? Yes, I believe it is.

By twisting the notion of climate denial on its head, Mura was able to find the strength to keep fighting a seemingly never-ending battle and making change. The piece also explores the difficulty of conversations regarding climate change with family members. The excellent metaphor that ties the work together makes Mura's message and call to action all the more poignant.

Another powerful piece submission in *Climate Denial* was the painting *The World I See* by 17-year-old Vidya Muthupillai. Featured on the cover of the issue, the image depicts a young woman staring off to the side with melancholy (Figure 12.2). Through a parting in her hair, the viewer sees that her mind is

FIGURE 12.2 "The World I See" by Vidya Muthupillai

depicted as the Earth. Muthupillai turned the prompt backward, showing the direct opposite of what a climate denier is.

"The bubble in which people perceive everything to be fine isn't the reality my generation and I live [in]—our world is on fire and we can't help but see it," she wrote in a paragraph accompanying the artwork. "Our activism seeks to shatter the barrier of willful ignorance and perception of safety because time is running out to fix climate change."

Climate Denial was received positively by many of the artists who submitted work. Muthupillai commented that she was "so glad to see that my work resonated with other people as well. [...] Thank you so much for doing everything you do to help shed light on concerns like climate denialism that have been such an overwhelming source of anxiety for so many youth today." Savanah Shadof wrote that "[Youth Think Climate] is a great opportunity and thank you for offering it!" Ethan Moeller told us, "Thank you so much for picking my piece to help spread the word about climate change!" An Instagram follower commented: "I truly enjoyed reading this first edition and I'm so excited for the next!! Thank you for making this happen!" These submitters and readers were deeply appreciative of the opportunity to have a place in which to share their work and reflect on the works of others.

Edition 2: Climate Justice

The second edition of the magazine was titled *Climate Justice* and was released in February 2021 (Figure 12.3). Included were 25 submissions from 12 US states and three from other countries (Canada, India, and Indonesia). As a climate organization, YTC wanted to put an emphasis on the fact that climate change is not just an issue of science (such as chemistry or biology), but also an issue of social justice.

The focus of the second edition was to bring into perspective how different people of different geographical, ethnic, and socioeconomic backgrounds experience the effects of climate change and the responses to climate change differently. This edition in particular seemed to contain works that were fearful, hopeless, and more negative about the future of humanity. However, it must be noted that many of these submissions were likely created during the period of time between the large-scale racial justice activism in the summer of 2020 and shortly after the 2020 presidential election and its ensuing turmoil. The submissions featured in the edition varied greatly but reflect much of the despair and anxiety experienced during this time. Some were academic analyses of climate justice through the problems of plastic waste and extreme weather, while others were poems that explored personal experiences with the subject matter.

FIGURE 12.3 "Climate Justice" Cover Image

Javy Polanco, a 15-year-old from New York, submitted the poem *Remember where you come from*, an example of a piece informed by international climate justice:

> Remember where you come from
> Not the city full of dreams
> Or the country full of corn,
> But the soil of the earth
> The forest of wild
> and the water of a lost sea

In Bolivia, lake Poopó is gone; evaporated
Skeletons of fish left
People forced to relocate

I wonder what other lakes we've lost; what other lakes we're losing
What other air we can't breathe fresh
What other forests we can't explore
What other people must suffer from it

So many decades of global prosperity
And barely a couple of centuries and it got lost
In factories and ignorance

Who is responsible,
Us as a species
Or us as a global society

What is our responsibility not only as citizens
But as animals of nature
What is our purpose to earth now that we know we're hurting her

What is our purpose in helping her heal?

Polanco's poem raises questions of responsibility in the climate crisis as well as provides a direct example of a country that disproportionately faces the severity of global warming in relation to the proportion of greenhouse gasses it emits. Climate justice, here gently intertwined with identity, brings a layer of depth to the poem and the uncomfortable truths it presents.

Climate justice was also portrayed in an untitled piece by Anika Kumar, a 17-year-old from Virginia (Figure 12.4). At the center of the image is an Earth on fire, with a mix of black-and-white geometric designs surrounding it. In the web of shapes, we see distinct images such as a police officer blinded by money, an LGBTQ+ pride flag, the women's gender symbol, a face mask, a raised Black Power fist, and a woman with a red handprint over her face to signify Missing and Murdered Indigenous Women.

Kumar commented that "Lots of nations face different types of issues and of course, they are all important, but one issue that everyone is facing globally is climate change. There needs to be more focus on climate change because we won't have a planet anymore if we don't start acting now." Kumar connects the struggle for the rights of these different identities to one intertwined with climate change, a key idea in the movement for climate justice.

FIGURE 12.4   "Untitled" by Anika Kumar

Edition 3: Earth Month

The third edition of the magazine was titled *Earth Month* and was released on April 22, 2021, in celebration of Earth Day (Figure 12.5). Within the edition, we included 45 submissions from three countries (India, Puerto Rico, and United Arab Emirates) and 16 US states.

APRIL 2021

EDITION 3

Youth Think Climate

EST. 2020

Your Submissions

Narratives, essays, poems, artwork, and photography on the beauty of our planet.

Earth Appreciation Interviews

Read about the impact nature has had on people during the pandemic and what actions you can take to help preserve it.

Cover artwork by Jasmine Parker - *Red Rock Falls* (pg 08)

EARTH MONTH EDITION

FIGURE 12.5 "Earth Month" Cover Image

The edition was divided into two distinct sections. The first was titled: "Appreciate Our Planet's Beauty." A majority of the submissions featured in this section were optimistic and appreciative. They tended to focus on Earth's beauty. The second section was titled: "Protect the Environment." This section hosted a majority of the more pessimistic submissions that expressed fear and disappointment.

Within both sections, a common inspiration is a reliance humans have on the resources that the Earth provides and our level of awareness of this relationship. One such piece is *Magnificence* by 14-year-old Juliana Verzi, which was included in the "Appreciate Our Planet's Beauty" section of the edition:

The sun's rays shine through the blue and white of the sky
aiding the natural processes of earth,
The radiant songs of birds filling the air
awakening creatures, large and small,
The vibrancy of flowers bringing color to the lush green valleys
the blossoming being's sweet nectar feeding the flamboyant butterflies and lively bees,
The snowflakes of varying shapes and sizes gleam as they fly through the air
hitting the ground and accumulating in soft, compressed fluffs until the cold passes,
The prismatic appearance of the blue sea from the surface
becomes a world abundant with life, thriving in the depths,
The dense jungles filled with exotic species
creating a multicolored ecosystem, vivid and radiant,
The savannahs with their golden grass,
the home for some of the world's mightiest and most massive wildlife
The grasslands, bustling with life of all forms,
those who graze on the fields of green
The mountains of gray, sleek and majestic
stand tall above all, ominous but breathtaking
These are some of the many components contributing to the beauty of nature, of earth
We cannot take these thing for granted as they need our help to flourish
We must learn to protect them to preserve earth's magnificence

A new addition we included with *Earth Month* was asking submitters interview questions so that we could learn more about their backgrounds and the inspiration behind their work. The first three questions were mandatory. For the remainder, artists could choose to respond to as many as they wanted.

Required:

1 What is your favorite animal from your local region?
2 What do you appreciate about the Earth/nature?
3 What is your favorite outdoor activity?

Optional:

4 How has nature helped you during the pandemic? Or, how has your perception or relationship with nature changed during the pandemic?
5 What actions have you taken to preserve the beauty of the Earth? What are some actions other people can take?
6 What motivates you to take action against climate change or other environmental crises?
7 What environmental crisis are you most passionate about fighting? Why?
8 What overall advice can you provide for people who want to start taking action to protect our planet?
9 Do you have any recommendations for environmental podcasts, authors, leaders, and/or documentaries we should look out for?

Our initial thought was that giving submitters more space and guiding questions would help better explain their work seemed effective. In addition, we also included many of the artists' answers alongside their work. This provided a unique way for readers to learn about the artist's values and their work with the climate movement.

An example of this is the painting *What We Leave Behind* by 20-year-old Amy Yang (Figure 12.6). In a lush underwater ecosystem, a turtle is surrounded by the ghostly remains of plastic pollution.

By additionally asking Yang our follow-up questions, we were able to gain a detailed background for this remarkable work:

> The inspiration for this piece came when I was asked, "What connects all of our oceans?" The sad thing is, the thing all of our oceans have in common is plastic. It's the things we throw in the trash, toss out the car window, or leave behind for someone else to clean up that will be the consequences that outlive us. The ones who pay the price, though, are the wildlife who will suffocate from our carelessness. I hope for this piece to encourage everyone to be aware of what they're buying, where it came from, and where it ultimately ends up. The legacy that humanity leaves behind can be more than just post-consumer waste.

Edition 4: Sustainability

The initial plan for YTC when the magazine started in the early months of 2020 was to only publish three editions to form a three-part series. With our fourth edition, we surpassed this original outline and are planning to continue. The fourth edition, titled *Sustainability*, included 23 submissions from nine different US states and five different countries (China, Germany, India, South

FIGURE 12.6 "What We Leave Behind" by Amy Yang

Africa, and Thailand) (Figure 12.7). Many of the pieces dealt with the future and how society can develop a sustainable way of life before catastrophic damage harms the Earth irreparably.

In the essay *Steering the Industry that Powers the World Through Sustainability*, 16-year-old Marlene Mostert from South Africa argues that large-scale global change is needed to achieve sustainability and protect the Earth.

Sustainability consists of three main pillars—social, economic, and environmental aspects. When we investigate the future of energy and fuel that powers our new-age civilization, we need to consider all aspects to find a solution.

Many activists around the globe demand that we end the use of fossil fuels and other non-renewable energy sources. I would love to believe that it is as simple as that because it would heal our Earth quickly. Governments around the world depend on such commodities to fund their financial growth, which directly influences the state of the country and its resources. Non-renewable

FIGURE 12.7 "Sustainability" Cover Image

energies can be replaced by innovative technology such as biofuels produced by waste or green energy solutions such as wind power, solar power, and others; however, the downfall of this transition is that it directly impacts the economy. In the long term, the investment in clean technology will pay off—but the paced greed of the powers will not decline a thriving commodity; therefore, the demand thereof will not be encouraged to decrease despite the fact it might just speed up the sixth mass extinction.

My take on this matter is that to choose a sustainable future in the industry that powers the world is not by protesting current management—there are endless loopholes people in powerful positions can jump through. Rather, we must reinvent the global civilization that has been systematically destroying our planet for decades. It not only affects our ecosystems which are integrated to sustain life on Earth but also affects the rural communities that have been exploited of their resources, resulting in unjust living standards and poor econ- omies. The resources we choose today do not only fail the environment, they fail all of us. The system is the way it is and has been manufactured that way through generations. The system is power steered by money. Money drives the world.

Creating great change is by steering the industry into a position where using sustainable technologies becomes the commodity. Green is the new rich. After all, what are numbers on a bank system going to achieve when the Earth starts to drown or suffocate by its own blazes?

We are inventors and problem solvers; we are creative and powerful, and mindful entrepreneurs. We are possibly the most advanced generations the world will ever know. It is up to us to preserve Earth for human civilization to co-exist with the biosphere and for future generations to come. Sustainability means that we will not compromise the needs of future generations to come, to create the impact we need to drastically reduce the rate of climate change, we need to form a global network to work together to introduce long-term solu- tions to achieve justice in the workplace, justice for the economy, and justice for the Earth.

As noted earlier, many of these pieces present a future safe from the climate crisis as within reach. Mostert does not tell the reader a world like this is im- possible; she tells them that we have the capabilities to actually make change.

On a smaller yet still critical level, the piece *Sustainable Culinary* by 20-year- old Californian Akio Goto examines sustainability within his food. In his words:

I try to promote a plant-based lifestyle and diet through the art of cook- ing. Growing up in a first-generation Japanese family, I understand the cul- tural importance of food in culture and communities. I try to retain the authenticity of communal eating by recreating flavors and cooking styles of traditional Japanese cooking in a more sustainable, vegan style. I hope to remind Japanese members of tastes they may have missed living in America while providing a sense of peace knowing their food is healthy for the planet, themselves, and to animals' well-being. The dishes [in Figures 12.8 and 12.9] are a vegan osetchi. Osetchi-ryori is a meal prepared to celebrate the New Year, typically prepared in advance so all people, including chefs, have a day off on New Year's. Every year, I enjoyed making osetchi with my family as

FIGURE 12.8   Image of Japanese Food

FIGURE 12.9   Close-up Image of Japanese Food

a way to retain culture and maintain family tradition. This year, I was able to convince my family to make our osetchi vegan and was inspired by everyone's openness to the improvised dishes. As a family, all we want is to be supportive of each other and enjoy time well spent while eating good food. I was proud to make a meal satisfying everyone's desires for Japanese food, enjoying each other's company, and doing so in a sustainable way.

This piece is unique because of its backstory and personal significance. Goto is able to connect with his family and culture through cooking, and in a deliberate way that he can ensure the planet is not harmed in the process. Works like these, which boldly present lived experience, bring a sense of humanity to the magazine. It informs readers that the artists are young people like them, with similar interests and motivations. Hopefully, readers come away from the pieces and feel not just empowered to make change, but that it is possible for them to accomplish it.

Bringing Climate Change Discussion into the Classroom

YTC believes that education and exposure to climate justice and climate activism, specifically in the context of youth, are critical to sustaining a dedicated and diverse climate movement. Our magazine model could be replicated within classrooms to facilitate discussion of various themes and ideas of the climate crisis. Students can tie in their own experiences to write personal-narrative style papers, produce short films, create artwork, and write fiction or poetry. Classes can then collaborate in the creation of a culminating magazine or exhibit that features these different pieces, assigning groups to organize and facilitate different areas or steps of the process. Along the way, teachers can lead class discussions on climate change.

Reading and critically engaging with the work of other young people during an era of climate crisis can provide several different benefits to a classroom. First, these types of artworks may motivate a student to take action, or even begin to change the way they view the climate movement and the role they play. Second, students may feel comforted in knowing that others have similar fears, questions, and emotions regarding the future of the planet. The end product (whether it be a digital or physical publication, exhibit, website, video, or something else entirely) can be distributed or shared with other classes, school staff, and parents to facilitate community-wide climate discussion. Students can also share their work on sites associated with youth publications so that they can connect with audiences beyond the classroom. The Climate Action page of the website Youth Voices *youthvoices.live/category/environment-and-world/ climate action*, for example, is a site where people can post their thoughts about global climate issues and solutions. Eventually, these projects, which capture knowledge and feelings from a specific moment in time, could be used to examine how thoughts on changes in the climate change movement over the years.

A necessary prerequisite to a project like this, however, must be a thorough climate education. Although there may be slight variances in the level of content matter for different age groups, students must be informed about the climate crisis from several different angles, including science, which may examine global warming and greenhouse gasses; social justice, which may study climate and environmental justice and Indigenous knowledge; and the humanities, which may analyze how the climate change is influencing societies, arts, and literature. Baseline knowledge not only provides students with the necessary information and skills for living in the 21st century but also allows them to think critically about the issue.

Teachers could engage students in studying, researching, and writing about climate change by simply giving them the freedom to explore (scientifically, personally, or fictitiously) a relevant sub-topic in any sort of medium. Climate change is a broad, multi-faceted issue with decades of history. Thus, students should ideally be able to choose something they are interested in to study, write about, or design. A personal connection to their work builds not an enjoyable creative process, but also encourages deeper self-reflection and effort.

For elementary school students, teachers or community members could come into classrooms and read climate-themed books to prompt further engagement, while middle or high school teachers may want to lead a climate-related club in which creation of art is a unit or activity, both of which may foster audiences beyond the classroom.

Conclusion

YTC is indebted to the work of dozens of young artists living in a world with an uncertain future. The magazine has been able to continue presenting artwork for several years because there are few spaces in which youth from across the world can come together and process the climate crisis and its impacts. As a result, we have garnered and fostered a unique community of hope and resilience on the online world. We hope to see projects that use YTC's model as a starting point for unique creative projects in classrooms, schools, community centers, and local and international organizations. Activism takes many forms; we hope that our work has given its contributors and readers hope and motivation to fight for true change. Whether this manifests itself as protesting or painting, they are declaring themselves the inheritors of the planet, ready to face what has long been ignored.